Switched Mode Power Supplies:
Design and Construction

ELECTRONIC & ELECTRICAL ENGINEERING RESEARCH STUDIES

ELECTRICAL ENERGY SERIES

Series Editor: **Dr H. W. Whittington**
The University of Edinburgh, UK

1. Switched Mode Power Supplies: Design and Construction
   H. W. Whittington, B. W. Flynn *and* **D. E. Macpherson**

Switched Mode Power Supplies: Design and Construction

H. W. Whittington

B. W. Flynn

and

D. E. Macpherson

The University of Edinburgh, UK

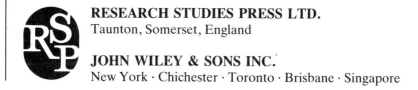

RESEARCH STUDIES PRESS LTD.
Taunton, Somerset, England

JOHN WILEY & SONS INC.
New York · Chichester · Toronto · Brisbane · Singapore

RESEARCH STUDIES PRESS LTD.
24 Belvedere Road, Taunton, Somerset, England TA1 1HD

Marketing and Distribution:

Australia and New Zealand:
JACARANDA WILEY LTD.
GPO Box 859, Brisbane, Queensland 4001, Australia

Canada:
JOHN WILEY & SONS CANADA LIMITED
22 Worcester Road, Rexdale, Ontario, Canada

Europe, Africa, Middle East and Japan:
JOHN WILEY & SONS LIMITED
Baffins Lane, Chichester, West Sussex, England

North and South America:
JOHN WILEY & SONS INC.
605 Third Avenue, New York, NY 10158, USA

South East Asia:
JOHN WILEY & SONS (SEA) PTE LTD
37 Jalan Pemimpin #05-04
Block B Union Industrial Building, Singapore 2057

Library of Congress Cataloging in Publication Data

Available.

British Library Cataloguing in Publication Data

A catalogue record for this book is
available from the British Library

   ISBN 0 86380 123 4 (Research Studies Press Ltd.)
   ISBN 0 471 93346 5 (John Wiley & Sons Inc.)

Printed in Great Britain by SRP Ltd., Exeter

EDITORIAL FOREWORD

This book is one of a series which is intended to cover recent developments in electrical energy conversion technology and applications.

Subjects covered in the series will include, but will not be limited to, the generation of electricity by novel or improved techniques.

H.W.Whittington
February 1992

This book is a support text for an 8-hour video course entitled Switched Mode Power Supplies. The video is available from EUMOS Ltd, University of Edinburgh, 16 Buccleuch Place, Edinburgh, United Kingdom, EH8 9LN. Tel: 031 650 3473, Fax: 031 662 4061. (Price on application)

CONTENTS

SECTION 1

INTRODUCTION & BASIC CIRCUITS

SECTION 2

COMPONENTS FOR SMPS

SECTION 3

SMPS CONTROL & DRIVER CIRCUITS

SECTION 4

PRACTICAL SMPS DESIGN CONSIDERATIONS

SECTION 1

INTRODUCTION
AND BASIC CIRCUITS

1.1   INTRODUCTION

All electronic circuits need a supply of power. For low power consumption units, often a battery, even a solar cell, will suffice. However, for most applications the power consumption is such that an electronic power supply must be used.

Over the past 10 years, there have been significant changes in the design of power supplies. The most important of these has been the widespread change from linear power supplies to those which operate on a switching basis - so-called Switched Mode Power Supplies (SMPS). The principal reason for the move to SMPS is their much greater efficiency - typically 80-90% as opposed to 30-40% for linear units. This greatly reduces the cooling requirements, and allows a much higher power density.

The concept of high frequency switching of transistors to provide a controllable d.c. output has been around for some time. What has allowed the widespread adoption of SMPS technology has been the availability of a range of suitable active and passive components. The advent of MOSFETs with high power rating has been a particularly important advance, together with the availability of high-speed diodes and improved magnetic materials. Now circuits can be designed to operate at switching frequencies into the megahertz range, with consequent reductions in cost and in volume of the power supply.

The market for SMPS in Europe in 1990 is estimated at 2 billion dollars, 80% of which will be designed in-house by the manufacturers of the host electronics. Paradoxically, despite this large market for power supplies, there are relatively few experienced power supply design engineers.

The power supply designer has to have a working knowledge of a wide range of skills:

- the design of power electronic switching circuits,

- the design of wound components,

- the understanding of control theory and its application to SMPS,

- the design for electromagnetic compatibility (EMC),

- an appreciation of the thermal management within SMPS.

The group at which this book is targeted includes SMPS designer engineers, equipment designers, and those responsible for the procurement of SMPS in user companies. The aim is to enable engineers to take advantage of recent developments in components and in circuit topologies, as well as covering more fundamental aspects of SMPS technology. Although the mathematical treatment, where used, is rigorous, it is kept to a minimum, the approach adopted throughout being essentially practical, with design "rules-of-thumb" introduced where appropriate.

1.2 SPECIFICATION OF SMPS

Any electronic system requires a power source in order to function correctly, yet the power supply is normally the most neglected part of the complete system. It is often hastily designed (or bought in) after the rest of the system is complete, and placed in the leftover space, which is often too small and has inadequate ventilation for cooling purposes.

System designers naturally wish to avoid any system faults occurring due to the power supply, so the specifications for the power supply normally include large safety margins. As the power supply will have its own safety margins, it is often grossly over-specified, and as a result is considerably larger, heavier and more expensive than is necessary.

When the systems designer is specifying the power supply, he will initially consider the required voltage, together with the maximum current. However, voltage and current ratings are only the beginning of a long list of parameters which need to be specified.

Number of Outputs

Typically, +5 volts is required for logic circuits, but frequently other voltage levels are required as well, possibly including negative voltages. Each output should have a complete specification.

Current Rating

Each output should have a maximum steady state current rating. In addition, any possible current spikes or surges (eg. starting disc drives) should be fully specified, preferably with a current waveform diagram.

Input Voltage

The input voltage must be specified, whether a.c. or d.c., together with the voltage range.

Isolation

In many applications, electrical isolation is required between the input and output, and between outputs, necessitating the use of a transformer in the power supply circuit. However, in some applications isolation is not necessary, and a smaller, cheaper power supply may be possible by choosing a design which does not incorporate an isolating transformer.

Ripple Voltage

This is normally expressed as a peak-to-peak voltage, at a fundamental frequency; however, it is sometimes given as a percentage of the nominal output voltage, and sometimes as an r.m.s. voltage. The ripple content can be reduced either by increasing the filter capacity, or by increasing the ripple frequency.

Regulation

The output voltage of a power supply is affected by:

(a) variations in the input voltage
(b) variations in the load current
(c) variations in temperature.

A regulated power supply has a feedback circuit to compensate for such changes, and keep the output voltage within specified limits. The compensation provided may be specified separately as Load Regulation, Line Regulation and Temperature Regulation.

(a) Line Regulation

Line regulation is a measure of the effect of changes in the input voltage on the output voltage.

There is a variety of methods used to specify line regulation; however, a common definition is:

$$Line\ regulation\ =\ \frac{\delta V_o}{V_o}\ \times\ \frac{1}{\delta V_{in}}\ \times\ 100\ \ \%\ /\ input\ volt$$

It should be noted that this definition implies a linear relationship between the input and output (ie. that a change in the input voltage will produce a proportional change in the output voltage), which often does not exist. For this reason, many manufacturers specify an input voltage range for which the power supply output will stay within specified limits.

(b) Load Regulation

Load regulation is the percentage change in the steady state output voltage when the load current is increased from zero to the fully rated current, or more commonly from 50% to 100% fully rated current (this does not take into account any transient effects which may occur).

Load regulation is defined as:

$$Load\ Regulation\ =\ \frac{(V_{NL}\ -\ V_{FL})}{V_{nom}}\ \times 100\ \%$$

where: V_{NL} =   output voltage on no-load

           V_{FL} =   output voltage on full load

           V_{nom} =   nominal output voltage

Load regulation is often quoted from 50% to 100% loading, rather than 0% to 100% loading.

(c) Temperature Regulation

Again, there are several different methods of quantifying the effect of temperature on the output voltage, but many power supply manufacturers specify a Temperature Coefficient, TC, where:

$$TC = \frac{(V_{o(max)} - V_{o(min)})}{V_{o(nom)}} \times \frac{1}{(T_{max} - T_{min})} \times 100 \ \% / °C$$

where: $V_{o(max)}$ =   output voltage at max. rated temperature

           $V_{o(min)}$ =   output voltage at min. rated temperature

           $V_{o(nom)}$ =   nominal output voltage

           T_{max} =   max. rated temperature

           T_{min} =   min. rated temperature

Many power supply manufacturers specify only an overall percentage regulation, together with limits for input voltage, output current, and temperature.

$$Regulation = \frac{(V_{max} - V_{min})}{V_{nom}} \times 100 \ \%$$

However, this figure is very much a worst case, as it is very unlikely that the extremes of load, input voltage and temperature will all occur simultaneously.

Transient Response

An often neglected power supply parameter is the response to sudden changes in load current. If the full load current is suddenly switched on when the power supply is very lightly loaded, the output voltage will dip temporarily below the minimum allowable steady state voltage (as set by the specified regulation). Conversely, if the load is switched off, the output voltage will rise temporarily above the maximum steady state voltage.

Transient response is normally specified as:

$$\frac{V_{dev}}{V_{nom}} \times 100 \% \quad \text{for } 100 \% \; load \; change$$

(V_{dev} = maximum deviation from nominal output voltage)

together with a "Recovery Time", the time it takes for the output to return within the specified regulation limits.

The transient response is affected by the type of power supply circuit used, the feedback loop and the output filter.

Efficiency

An inefficient power supply has two major disadvantages:

(a) Energy is wasted, which is particularly important when fed from a battery.

(b) Large heatsinks and good ventilation are required, which add to the size and weight of the power supply.

Protection

Power supplies can have many different forms of protection, but the most common are:

(a) Overvoltage
It is important to protect the load against an overvoltage at the output. It is normal to simply shut down the power supply if an overvoltage is detected (apart from transients). Alternatively, a "crowbar" thyristor may be used, where a thyristor is triggered if the voltage rises above a preset limit, putting a short circuit across the output, and hence clamping the output voltage low, with the current controlled by the current limit circuit.

(b) Overcurrent
Most power supplies have some form of current limit, such that if the load current rises above a preset level, the output voltage reduces to limit the current to a safe level.

(c) Short Circuit
This may be covered by the overcurrent protection, but a separate circuit is often used to shut down the power supply if a short circuit is seen across the output terminals.

(d) Inrush Current
SMPS normally have a large d.c. smoothing capacitor near the input, which is likely to cause a very high current spike at initial switch-on. Most SMPS incorporate some form of current limiting circuitry to keep this inrush current to a minimum.

Electromagnetic Interference (EMI)

This is not normally a problem with linear or low frequency (< 20 kHz) switched mode power supplies, but the problems associated with EMI increase with frequency. Adequate filtering of input and output leads should be provided as a minimum.

Hold-Up Time

It is highly desirable that the SMPS should continue to provide an output voltage within specification through short interruptions to the input lasting, typically, one or two cycles at 50 Hz.

Temperature Range

It should be noted that the ambient temperature is the temperature INSIDE the cabinet/box in which the system and the power supply are housed, which is likely to be considerably higher than the temperature outside. Allowance should also be made for the temperature rise due to the power dissipated by the power supply.

Physical Dimensions

The physical size and weight of the power supply are often the limiting factor in the choice of power supply, as it is usually situated in the leftover space. Recent advances in high frequency (> 100 kHz) switched mode power supplies have resulted in considerable reductions in size.

Certification

Many countries have their own regulations for SMPS. For example, in Continental Europe, one of the most important bodies is West Germany's *Verband Deutscher Electrotecniker* (VDE) which sets the pace for national testing agencies, with many of its requirements being very tight indeed.

Of great significance to the power supply user are the various electromagnetic interference (EMI) regulations. Although VDE has, for some time, had EMI specifications, it is relatively recently that the USA issued its own specifications. Most of the end products that power supplies are used in must pass Underwriters

Laboratory (UL) tests for use in the US, and Canadian Standards Association (CSA) tests for use in Canada. Approval of the end product by these organisations is easier to obtain if the components - including power supplies - have previously been recognised or certified.

It should be noted that, although UL and CSA approval is required for most electrical products sold for use in the US and Canada, it is of little value elsewhere in the world.

The relevant IEC (International Electrotechnical Commission) safety standard is IEC 380, which requires a 3750 volt a.c. withstand between input and output, necessitating a minimum spacing of 8 mm between primary and secondary circuits, and 3 mm creepage between live parts to dead metal. This is considerably more strict than the relevant US standards.

The relevant EMI (Electromagnetic Interference) standards are IEC 478 part 3, the German VDE 0871, and BS 800 (1983). British Standards have also produced a Code of Practice CP 1006 "General Aspects of R.I. Suppression".

Economics

As usual, the cost of the power supply is often the overriding consideration. Frequently, system designers underestimate (or ignore altogether) the cost of the power supply, while expecting very tight specifications to be met.

1.3 RECTIFIER CIRCUITS

The block diagram of a typical SMPS is shown below. This section discusses the design of the two rectifiers included in most power supplies.

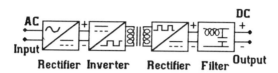

SMPS block diagram

There are two circuits used for converting single phase a.c. to d.c., the bridge rectifier and the centre-tapped transformer rectifier (shown below). They both have the same output waveshape, shown for both sinusoidal input and square wave inputs.

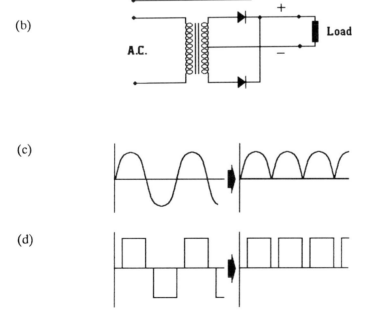

(a) Bridge rectifier (b) Centre-tapped transformer rectifier
(c) Waveform for sinusoidal input (e) Waveform for square wave input

Input Rectifier

This rectifier converts the a.c. mains voltage to high voltage d.c. A bridge rectifier is normally used, to avoid the inclusion of a low frequency transformer, which would add considerably to the size and weight (and cost) of the power supply.

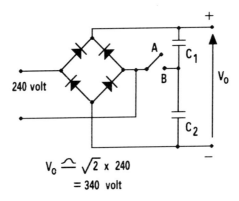

Input rectifier for 240 or 110 volt a.c. input

The inclusion of the switch and centre-tapped capacitor network allows the power supply to operate either from a 220-240 volt a.c. input, or a 110 volt a.c. input.

In position A, the rectifier will operate as a standard diode bridge, with a peak output voltage V_o:

$$V_o = \sqrt{2} \times 240$$
$$\approx 340 \quad volts$$

The capacitors will smooth the V_o as shown below.

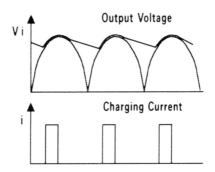

Input rectifier voltage and current waveforms

It can be shown that the value of capacitance is given by:

$$C = \frac{power}{V_{max}^2 - V_{min}^2} \times \frac{1}{f}$$

where:

V_{max} = peak d.c. voltage
V_{min} = minimum d.c. voltage
f = a.c. mains frequency (50 Hz)

The charging current is also shown. It can be seen that this is a narrow current spike, which will become narrower and higher if V_{min} is increased by increasing the smoothing capacitor.

It can be shown that the magnitude of the charging current is given by:

$$i_{peak} = \frac{C}{t_c} \times (V_{max} - V_{min})$$

where:

$$t_c = \frac{1}{2\pi f} \times \cos^{-1}\left(\frac{V_{min}}{V_{max}}\right)$$

This current spike can introduce harmonics onto the mains supply, which may need filtering out. Legislation will shortly be introduced governing the waveshape of current drawn from the mains.

If a 110 volt input is used, then the switch should be moved to position B. In this case the upper capacitor will be charged during the positive half cycle, and the lower capacitor during the negative half cycle, giving a peak output voltage:

$$V_o = 2 \times \sqrt{2} \times 110 \quad volts$$

$$\approx 311 \ volts$$

which is only 9% less than from a 240 volt supply, with the switch in position A.

Output Rectifier

The output rectifier is normally a centre-tapped transformer rectifier. In a bridge rectifier two diodes are always conducting, which would add considerably to the losses, particularly for low voltage outputs (e.g. 5 volt supplies), due to the relatively high diode on-state voltage. However in the centre-tapped transformer rectifier there is only ever one diode conducting at any instant. Most SMPS include a high frequency transformer for isolation, so an extra transformer is not required.

It should be noted that the diodes should be rated for twice the peak d.c. voltage.

1.4 BASIC CONVERTER CIRCUITS

Many different SMPS circuit topologies have been developed, most of which are derivatives of the following basic circuits:

(a) buck

(b) flyback (sometimes called buck-boost)

(c) boost

BUCK REGULATOR

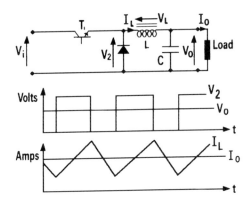

Buck regulator with associated voltage and current waveforms

In the basic buck circuit, transistor T_1 is switched at high frequency (20 kHz → 1 MHz) to produce a chopped output voltage V_2. This is then filtered by the L-C circuit to produce a smooth load voltage V_o. The output voltage can be controlled by varying the mark:space ratio of V_2.

The average output voltage is given by:

$$\frac{V_o}{V_i} = \frac{t_{on}}{T}$$

$$= D$$

where: T = total period = $\dfrac{1}{frequency}$

t_{on} = transistor on time

D = transistor duty ratio = $\dfrac{t_{on}}{T}$

Thus in the ideal case the output voltage is independent of load.

However, in a realistic circuit there will be losses associated with:

(a) transistor "on" state voltage drop

(b) transistor switching losses

(c) diode forward voltage drop

(d) inductor effective resistance,

all of which increase the dependency of V_o on load.

When transistor T_1 is switched on, $V_2 = V_i$, so the voltage across the inductor is:

$$V_L = L \cdot \frac{di}{dt} = V_i - V_o$$

So I_L increases linearly.

When transistor T_1 is switched off, the current through the inductor L cannot instantaneously fall to zero, so the "flywheel" diode is included to provide a return path for the current to circulate through the load. During this period, $V_2 = 0$, so:

$$V_L = L \cdot \frac{di}{dt} = -V_o$$

So I_L decreases linearly.

The peak-to-peak current ripple is:

$$\Delta I = t_{on} \cdot \left(\frac{V_i - V_o}{L} \right)$$

$$= \left(\frac{V_i \cdot T}{L} \right) \cdot D \cdot (1 - D)$$

which reaches a maximum when $D = 0.5$

The maximum current is:

$$I_{max} = \frac{V_o}{R} + \frac{\Delta I}{2}$$

The minimum current is:

$$I_{min} = \frac{V_o}{R} - \frac{\Delta I}{2}$$

Discontinuous Mode Operation

The previous section analysed the buck converter for the *continuous current* mode, ie. when the inductor current never falls to zero in the cycle. However, many switched mode power supplies are designed to operate in the *discontinuous current mode*, where for a proportion of each cycle the inductor current is zero.

If the load on a buck regulator is slowly reduced, eventually $I_{min} = 0$ (case (ii) below). If the load is reduced still further, the inductor current is zero for a period, and the circuit is operating in the discontinuous mode (case (iii)).

(i)

(ii)

(iii)

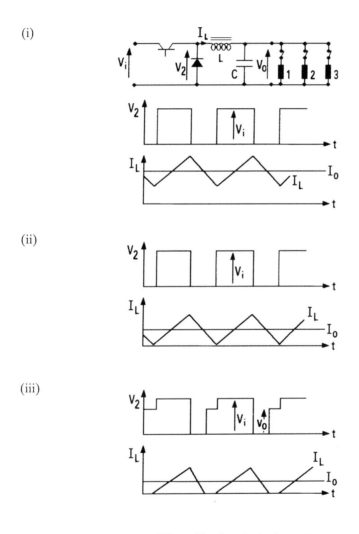

Effect of load on the buck regulator
(i)   Continuous inductor current (full load)
(ii) Boundary between continuous and discontinuous inductor current
(iii) Discontinuous inductor current (light load)

For t_{on} , inductor voltage $V_L = V_i - V_o$.

For the period the diode is conducting $(D_1.T)$, the inductor voltage $V_L = V_o$.

Since the average voltage across an inductor over a complete cycle is zero (assuming steady state conditions),

$$(V_i - V_o).D.T = V_o.D_1.T$$

Therefore:

$$\frac{V_o}{V_i} = \frac{D}{D + D_1}$$

The ripple current is:

$$\Delta I = \frac{(V_i - V_o).D.T}{L}$$

$$= \frac{V_o.D_1.T}{L}$$

The output current is:

$$I_o = \left(\frac{\Delta I}{2}\right) . \left[\frac{(D.T + D_1.T)}{T}\right]$$

$$= \frac{V_o}{R}$$

Combining the above:

$$\frac{V_o}{V_i} = \frac{2.D}{D + \sqrt{D^2 + (8.L/R.T)}}$$

Choice of Filter Inductor

At the boundary condition between continuous and discontinuous mode of operation,

$$I = \frac{V_o}{R} = \frac{\Delta I}{2}$$

Therefore:

$$L = \tfrac{1}{2}.R.T.(1 - D)$$

The value of L decides whether the circuit will operate in the continuous mode or discontinuous mode.

For continuous mode operation:

$$L > \tfrac{1}{2}.R.T.(1 - D)$$

for all values of R and D .

For discontinuous mode operation:

$$L < \tfrac{1}{2}.R.T.(1 - D)$$

for all values of R and D.

Choice of Filter Capacitor

The output filter capacitor is chosen to minimise the output ripple voltage, ΔV_o.

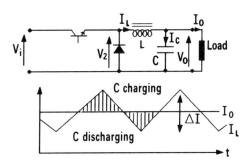

Charging and discharging the filter capacitor of a buck regulator

The change in charge ΔQ of the output capacitor is represented by the shaded area in the diagram. Thus:

$$\Delta Q = \frac{1}{2} . \frac{\Delta I}{2} . \frac{T}{2}$$

$$= \frac{T . \Delta I}{8}$$

Thus:

$$\Delta V_o = \frac{\Delta Q}{C}$$

$$= \frac{T . \Delta I}{8 C}$$

$$= \frac{V_o . T^2}{8 L C} . (1 - D)$$

or:

$$C = \frac{V_o}{\Delta V_o} . \frac{T^2}{8 L} . (1 - D)$$

However, in most practical circuits, the output ripple voltage is more likely to be caused by the ripple current through the capacitor ESR (equivalent series resistance).

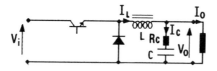

Buck regulator including capacitor ESR

The peak-to-peak ripple voltage is thus given by:

$$\Delta V_o = \Delta I \cdot R_c$$

where:

R_c = capacitor ESR

$$\Delta I = \frac{t_{on} \cdot (V_i - V_o)}{L}$$

It is, therefore, important to choose a filter capacitor with a low ESR value.

FLYBACK REGULATOR

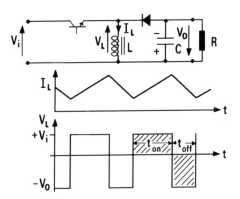

Flyback regulator with inductor current and voltage waveforms

In the basic flyback regulator shown above, energy is stored in inductor L during the on-time of transistor T_1. When T_1 is switched off, the voltage across L reverses as the inductor transfers the stored energy ($= \frac{1}{2}Li^2$) to the smoothing capacitor and to the load (this is the "flyback" period). Note that the output voltage is the opposite polarity to the input.

The basic equation for an inductor is:

$$V_L = L.\frac{di}{dt}$$

Integrating both sides over one complete cycle:

$$\int V_L.dt = \int L.di = L.(I_2 - I_1)$$

where I_1 is the current at the start of the cycle, and I_2 at the end of the cycle. Under steady state conditions, the current through the inductor at the start of a cycle equals the current at the end of the cycle; therefore, over one complete cycle:

$$\int V_L.dt = 0$$

ie. the *average* voltage across the inductor over one complete cycle is zero.

During the on period, the inductor voltage $V_L = V_i$, while during the off period, $V_L = -V_o$

Therefore:

$$V_i.t_{on} - V_o.(T - t_{on}) = 0$$

Therefore:

$$\frac{V_o}{V_i} = \frac{t_{on}}{T - t_{on}}$$

$$= \frac{D}{1 - D}$$

where D = transistor duty ratio.

The d.c. gain (V_o/V_i) is affected considerably by the effective inductor resistance, particularly as D approaches unity. Taking this into account:

$$\frac{V_o}{V_i} = \left(\frac{D}{1 - D} \right) \cdot \left[\frac{(1 - D)^2 . R}{(1 - D)^2 . R + r_L} \right]$$

resulting in a maximum of:

$$\frac{V_o}{V_i} = \frac{1 - \sqrt{r_L/(r_L + R)}}{2 . \sqrt{r_L/(r_L + R)}}$$

Other losses which have been neglected are the transistor on-state voltage drop, the forward volt drop across the diode, and the transistor switching losses. However, the formulae as presented are usually adequate as a first approximation.

The peak-to-peak current ripple is:

$$\Delta I = t_{on} \cdot \left(\frac{V_i}{L} \right)$$

$$= D \cdot T \cdot \left(\frac{V_i}{L} \right)$$

At the boundary condition between continuous and discontinuous mode of operation, $I_{min} = 0$, and the average current into the load is:

$$I = \frac{V_o}{R}$$

$$= \frac{\Delta I}{2} \cdot \frac{t_{off}}{T}$$

$$= \frac{D.T.V_i}{L} \cdot \frac{(1 - D).T}{T} \cdot \frac{1}{2}$$

Thus:

$$L = \tfrac{1}{2} . R . T . (1 - D)^2$$

Therefore, for continuous mode operation:

$$L > \tfrac{1}{2} . R . T . (1 - D)^2$$

for all likely values of R and D.

- 20 -

Discontinuous Mode Operation

If load is slowly reduced, the flyback regulator will eventually be operating in the discontinuous mode.

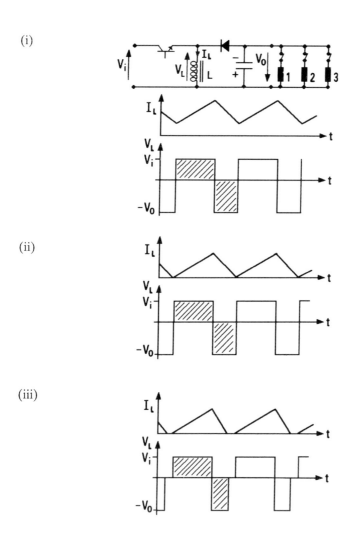

Effect of load on the flyback regulator
(i) Continuous inductor curent (full load)
(ii) Boundary between continuous and discontinuous inductor current
(iii) Discontinuous inductor current (light load)

The average voltage across an inductor over one complete cycle equals zero; therefore:

$$V_i.D.T = V_o.D_1.T$$

Hence:

$$\frac{V_o}{V_i} = \frac{D}{D_1}$$

In a similar fashion as earlier, it can be shown that:

$$\frac{V_o}{V_i} = D.\left[\frac{R.T}{2L}\right]^{\frac{1}{2}}$$

Choice of Filter Capacitor

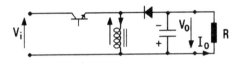

Flyback regulator

The ripple voltage is obtained as follows:
During the transistor on-time, the load current is supplied by the filter capacitor. Thus:

$$I = -C.\frac{dV_o}{dt} = \frac{V_o}{R}$$

or:

$$\frac{dV_o}{dt} = \frac{\Delta V_o}{\Delta t}$$

$$= -\frac{V_o}{RC}$$

$$\Delta t = t_{on} = D.T$$

so:

$$\frac{\Delta V_o}{V_o} = \frac{D.T}{R.C}$$

However, as with the buck regulator, the more important parameter for limiting the ripple voltage is usually the capacitor ESR.

Unlike the continuous mode buck regulator, the capacitor in the flyback regulator supplies *all* the load current during the period that the transistor is on. Therefore the capacitor in the flyback regulator sees a much higher ripple current, placing a more stringent ESR requirement on it.

BOOST REGULATOR

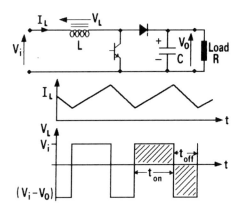

Boost regulator with inductor current and voltage waveforms

During the on-time of transistor T_1, the current builds up in inductor L due to the inductor voltage $V_L = V_i$. When T_1 is switched off, the voltage across L reverses ($V_L = V_i - V_o$) and adds to the input voltage, making the output voltage greater than the input voltage.

As before, the average voltage across the inductor over one complete cycle is zero, therefore:

$$V_i . t_{on} - (V_o - V_i) . t_{off} = 0$$

Therefore:

$$V_i . D . T = (V_o - V_i) . (1 - D) . T$$

And:

$$\frac{V_o}{V_i} = \frac{1}{1 - D}$$

where D = transistor duty ratio.

The d.c. gain (V_o / V_i) is affected considerably by the effective inductor resistance, particularly as D approaches unity. Taking this into account:

$$\frac{V_o}{V_i} = \left(\frac{1}{1 - D} \right) . \left[\frac{(1 - D)^2 . R}{(1 - D)^2 . R + r_L} \right]$$

resulting in a maximum:

$$\frac{V_o}{V_i} = \frac{1}{2} . \left[\frac{R}{r_L} \right]^{\frac{1}{2}}$$

The effect of inductor resistance on the d.c. gain is shown in the figure below.

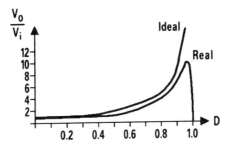

Effect of inductor resistance on the d.c. gain of a boost regulator

Other losses which have been neglected are the transistor on-state voltage drop, the forward volt drop across the diode, and the transistor switching losses. However, the formulae as presented are usually adequate as a first approximation.

The current ripple is:

$$\Delta I = \left(\frac{V_i}{L} \right) . t_{on}$$

$$= \left(\frac{V_i}{L} \right) . D.T$$

Input power = output power,

$$V_i . I_{L(av)} = V_o . I_{load}$$

Therefore:

$$I_{load} = \frac{V_o}{R} = I_{L(av)} . \frac{V_i}{V_o}$$

$$= I_{L(av)} . (1 - D)$$

At the boundary condition between continuous and discontinuous mode of operation,

$$I_{L(av)} = \frac{\Delta I}{2}$$

Therefore:

$$\frac{V_o}{R} = \frac{\Delta I}{2} . (1 - D) = \frac{V_i}{2.L} . D.T . (1 - D)$$

Therefore:

$$L = \tfrac{1}{2}.R.T.D. (1-D)^2$$

Thus, for continuous mode of operation:

$$L > \tfrac{1}{2}.R.T.D. (1-D)^2$$

for all likely values of R and D.

Discontinuous Mode of Operation

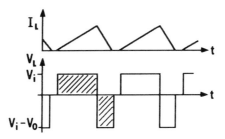

Inductor current and voltage waveforms in the discontinuous current mode

Again, the inductor volt-second relationship holds:

$$V_i.D.T = (V_o - V_i).D_1.T$$

So:

$$\frac{V_o}{V_i} = \frac{D + D_1}{D_1}$$

It can be shown that:

$$\frac{V_o}{V_i} = \frac{1 + \sqrt{1 + 2.D^2.T.R/L}}{2}$$

Choice of Filter Capacitor

As with the flyback regulator, the load is supplied by the filter capacitor during the transistor "on" period. Hence, the output ripple voltage is the same as for the flyback regulator, ie.

$$\frac{\Delta V_o}{V_o} = \frac{D.T}{R.C}$$

Again, the capacitor ESR is likely to be of paramount importance in the choice of capacitor.

SUMMARY OF BASIC CONVERTER CIRCUITS

Circuit Summary

Topology	Polarity V_o wrt V_i	Magnitude V_o wrt V_i
Buck	same polarity	step-down
Flyback	inverse polarity	step-up or step-down
Boost	same polarity	step-up

Continuous or Discontinuous Mode of Operation

(a) Continuous

In this mode, the inductor current, I_L , never falls to zero during any part of the switching cycle. To maintain the current, the inductance has to be considerably larger than is required with the discontinuous mode.

Although the inductor current is continuous, in the flyback and boost regulators the current into the regulator output stage (ie. the diode current) is discontinuous. However, in the buck regulator the inductor current is the current into the output stage, which is continuous and has a relatively small ripple value. The buck regulator is, therefore, easier to filter, and is the most popular switching configuration, particularly at high power levels.

Open Loop Regulation

The basic equations for continuous mode operation derived earlier are:
Buck:
$$\frac{V_o}{V_i} = D$$
Flyback:
$$\frac{V_o}{V_i} = \frac{D}{1 - D}$$
Boost:
$$\frac{V_o}{V_i} = \frac{1}{1 - D}$$

It can be seen from the above that the output does not depend on the load resistance R (taking a first order approximation and neglecting the inductor resistance r_L). Hence the open loop load regulation is very good. However, V_o does depend on the input voltage V_i , so the open loop line regulation is poor.

Closed Loop Response

The large inductor required for continuous mode operation, together with the filter capacitor, constitutes a 2nd order delay in the feedback control loop, resulting in poor closed loop response. This is covered in more detail in Section 3.2.

(b) Discontinuous

In this mode the inductor current falls to zero each cycle. This results in high inductor current peaks, placing an arduous duty on the switching transistor and the filter capacitor, as well as the inductor itself.

Open Loop Regulation

As shown earlier in this section, when a buck, flyback or boost regulator operates in the discontinuous mode the output voltage depends on the load resistance R. Hence the open loop load regulation is poor.

As in the continuous mode, the output voltage depends on the input voltage, so the open loop line regulation in the discontinuous mode is also poor.

Closed Loop Response

As the inductor starts each cycle with zero stored energy, it is possible for the control circuit to obtain any energy level and hence output current on a cycle-by-cycle basis. The inductor, therefore, has no effect on the small signal closed loop characteristic, leaving only the capacitor as the delay element in the loop. Regulators operating in the discontinuous mode, therefore, are very stable and have a very good closed loop response.

General

It can be seen from the above that each of the three types of regulator has very different open loop and closed loop characteristics, depending on whether operation is in the continuous or discontinuous current mode. It is important that a regulator designed for one mode of operation is not used in the other, as the different feedback characteristic is likely to lead to instability.

If the current in a continuous mode regulator is reduced below a minimum value (ie. a high value of R), discontinuous mode operation will result. Thus a continuous mode regulator should not be run on very light loading.

1.5 ISOLATED SWITCHED MODE POWER SUPPLIES

Most SMPS are required to have transformer-coupling between the input and the output(s). Transformer coupling provides the following advantages over the basic regulators so far described:

(1) The output is electrically isolated from the input. This is usually a requirement when operating from a 240 volt or 110 volt mains supply, to keep the mains voltages well apart from a low voltage load. It also allows the output to be earthed, if required.

(2) The transformer turns ratio can be chosen to give an output voltage widely different from the input voltage. The basic buck, flyback and boost circuits have the output voltage limited to within approximately a factor of 10 of the input voltage.

(3) With transformer coupling the polarity and step-up/step-down restrictions of the basic circuits no longer apply.

(4) By having more than one transformer secondary, multiple outputs at different voltage levels can be obtained.

However, the introduction of a transformer adds considerably to the size and weight of the SMPS, and introduces further losses into the circuit. In addition, the transformer leakage inductances may lead to severe voltage spikes in the circuit.

FLYBACK CONVERTER

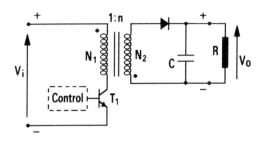

Flyback converter with isolation

This circuit operates in a very similar manner to the basic flyback regulator, but here the inductor has a secondary winding (or windings). Note that the wound component is an inductor with a secondary winding, combining the functions of both an inductor and a transformer. An inductor is an energy storage device, and as such requires an air gap in the magnetic circuit (it is not possible to store significant amounts of energy in the ferromagnetic part of the core). An ideal transformer directly couples energy between the primary circuit and the secondary, and does not store energy. In this circuit an air gap is required, as energy is stored in the device.

During the transistor on-time, current builds up in a linear manner in the primary circuit ($V_i = L. \ di/dt$), storing energy ($= \frac{1}{2}.L.i^2$) in the inductor. During this period, diode D_1 prevents any current flowing in the secondary.

During the transistor off-time, the energy stored in the inductor is released to the load.

It can be shown that for continuous mode operation:

$$\frac{V_o}{V_i} = \frac{n.D}{1 - D}$$

where $n = \dfrac{N_2}{N_1}$ = turns ratio

For continuous mode operation, the primary inductance should be:

$$L > \frac{R.T.(1 - D)^2}{2.n^2}$$

The (primary) current ripple is:

$$\Delta I = D.T.\left(\frac{V_i}{L}\right)$$

as with the basic flyback regulator.

The flyback converter is more commonly used in the discontinuous mode, due to:

(a) smaller inductor required

(b) better closed loop response

The disadvantages of discontinuous mode operation are:

(a) high peak transistor current (approx. twice continuous mode)

(b) large filter capacitor (due to high peak current)

In practice the isolated flyback converter has two very useful features:

(1) It is able to boost the input voltage (independent of the transformer turns ratio), making it very attractive as a low power EHT supply (eg. in television sets, computer monitors etc.).

(2) Because the wound component is the energy storage inductor, any secondary only requires a diode and a capacitor to produce an independent d.c. supply. Additionally it is possible to include an extra winding for the feedback signal, providing isolation in the feedback circuit, yet without introducing errors due to the volt drop across an output inductor. It is thus an attractive circuit when a number of independent outputs are required.

The isolated flyback converter is very common for powers up to about 200 watts.

FORWARD CONVERTER

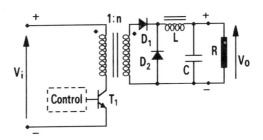

Basic forward converter

The forward converter is very similar to the basic buck regulator, but the addition of a transformer provides both isolation and the possibility of a very wide voltage range.

During the transistor on-time, diode D_1 is forward biased, and energy is transferred from the input to the load. During the off-time, D_1 is reverse biased and D_2 is forward biased to maintain continuous current in the output circuit.

In the circuit above, the primary voltage is always greater than or equal to zero, so the transformer core is likely to saturate. There are a number of schemes to reset the core, some of which include extra switching components. A common technique is to include a tertiary winding on the transformer core, as shown below, which conducts during the off period, resetting the core.

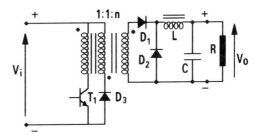

Forward converter with tertiary winding

During the on-time, D_3 is reverse biased. However, during the off period D_3 is forward biased putting V_i across the tertiary winding, which acts to reset the core. During this period, the tertiary winding induces a voltage of $-V_i$ across the primary, which adds to the input voltage to put $2.V_i$ across the transistor (if the tertiary winding has the same number of turns as the primary).

To avoid saturating the transformer, the applied volt-seconds must equal the demagnetising volt-seconds (ie. the average voltage across the primary must equal zero); therefore the duty cycle D must not exceed 0.5 (50 %). This results in poor transformer utilisation as energy is being transferred from the primary to the secondary only during the on period. If the turns ratio of the primary to the tertiary winding is increased (number primary turns > number tertiary turns), the duty ratio may exceed 0.5 , but the resulting voltage across the transistor during the off period will then be greater than $2.V_i$. When operating from a rectified mains voltage (typically 340 volt d.c.), this places a severe duty on the transistor.

The d.c. gain of the forward converter is:

$$\frac{V_o}{V_i} = n.D$$

The forward converter is normally used in the continuous mode, where the low current ripple does not place a heavy duty on the filter capacitor. However, in the continuous mode the closed loop response is poor (similar to the buck regulator) and can be difficult to stabilise. At high powers (>500 watts) the poor transformer utilisation results in a bulky design, and other circuits are likely to be preferred.

PUSH-PULL CONVERTER

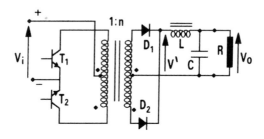

Push-pull converter

In the push-pull converter above, another buck-derived circuit, T_1 and T_2 conduct alternately, causing currents I_1 and I_2 respectively, and producing the transformer secondary voltage V_s as shown below. This is then rectified and filtered to produce a smooth d.c. voltage.

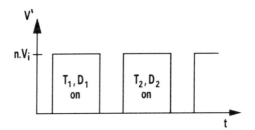

Rectified output voltage of a push-pull converter (before filter)

The duty ratio of each transistor must be limited to 0.5 (50 %), to prevent an effective short circuit of the d.c. supply (this will occur if both transistors conduct simultaneously).

Energy is transferred from the primary to the secondary during both halves of the cycle, so the transformer utilisation is much better than in the forward converter.

Care must be taken to ensure that each transistor conducts for exactly equal periods, or a d.c. voltage component will be applied to the transformer primary which will cause the transformer to saturate. Unequal transistor switching and storage times, and unequal transistor saturation voltages, can also cause the transformer to "walk" into saturation. A transformer with an air gap in the core is usually used to prevent saturation occurring.

The input voltage V_i is applied across half of the primary winding, so the voltage across the full primary is $2.V_i$. Hence the transistors need to have a voltage rating in excess of $2.V_i$.

The push-pull converter is essentially a buck-type regulator with isolation. The d.c. gain is:

$$\frac{V_o}{V_i} = 2.n.D$$

where: n = the transformer turns ratio
D = the duty ratio of each transistor.

As with all buck type regulators, when operated in the continuous mode (as they usually are) the output current ripple is very low, so the duty on the filter capacitor is low. However, in the continuous mode the closed loop response is poor, and the loop can be difficult to stabilise.

The push-pull converter has the following advantages:

(a) it is efficient,

(b) it is compact,

(c) it provides isolation between the input and output,

(d) it utilises the transformer well,

(e) unlike the half-bridge and the bridge converters, both transistors are referred to the same voltage (ie. the emitters are connected together), so they can be driven directly by the control circuit, without the need for isolating transformers in the base/gate drive.

Its disadvantages are:

(a) two switching transistors are required,

(b) the transistor voltage rating must be 2 x input voltage,

(c) the transistors must be on for exactly equal periods or a d.c. component will be applied to the transformer, which will cause the core to saturate.

HALF-BRIDGE CONVERTER

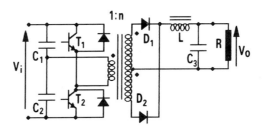

Half-bridge converter

In the half-bridge converter, capacitors C_1 and C_2 form a voltage divider of the input voltage V_i , the average voltage at the centre-point being $\frac{1}{2}V_i$. Transistors T_1 and T_2 conduct alternately, switching the primary voltage from $+\frac{1}{2}V_i$ to $-\frac{1}{2}V_i$. Thus the peak transistor voltage is V_i , half the value of the push-pull converter. However, for an equal power rating, the half-bridge transistors carry twice the current of the push-pull transistors.

The d.c. gain of the half-bridge is:

$$\frac{V_o}{V_i} = n.D$$

As with the push-pull converter, the duty ratio D of each transistor must be limited to 0.5, to prevent both transistors conducting simultaneously, and thus causing a short circuit across the supply voltage.

An advantage the half-bridge has over the push-pull is that capacitors C_1 and C_2 block any d.c. component to the transformer primary, reducing the risk of saturation of the iron due to unequal transistor conduction intervals.

As T_1 and T_2 are in series, the base drives cannot be controlled from a common reference voltage. Usually the base drive circuits have to be transformer coupled, which adds considerably to their complexity.

The half-bridge is a further adaptation of the basic buck regulator. It is popular in power ratings from 500 watts to 2000 watts.

BRIDGE CONVERTER

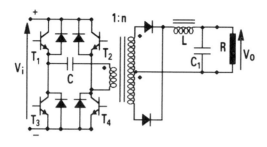

Full bridge converter

In the bridge converter, transistors T_1 and T_4 conduct together, then transistors T_2 and T_3, thus producing a square a.c. voltage waveform equal to $\pm V_i$ on the transformer primary.

The d.c. gain is:

$$\frac{V_o}{V_i} = 2.n.D$$

Capacitor C is included to block any d.c. component arising due to unequal conduction periods on the positive and negative half-cycles, which would lead to saturation of the transformer core. With the capacitor, any imbalance in the conduction periods merely produces a shift in the mean d.c. level of the waveform, so the transformer sees a balanced volt-seconds on the two half-cycles.

The bridge converter is generally used in high power applications (greater than 750 watts) where the half-bridge capacitors C_1 and C_2 become very large. (With power ratings in excess of 1kW, capacitor C is sometimes left out, as it also becomes very large.) The output power from a bridge converter is double that from a half-bridge with equally rated transistors. However, two extra transistors are required, and the base drive circuits are similarly complex, requiring transformer coupling.

SWITCHED MODE POWER SUPPLIES WITH MULTIPLE OUTPUTS

In many applications, more than one output is required, with each output likely to have different voltage and current specifications.

The basic buck, flyback and boost regulators are not suitable for multiple output applications. However, multiple outputs can be readily obtained using any of the converters which have an isolating transformer, by employing a separate secondary winding for each output, as shown in the forward converter below.

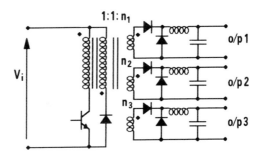

Forward converter with 3 outputs

In the above circuit, each output voltage will be determined by the appropriate turns ratio n_1, n_2 or n_3.

1.6 CIRCUIT DESIGN CONSIDERATIONS

CHOICE OF SWITCHING FREQUENCY

The choice of switching frequency depends on a compromise between size and efficiency. The size of the transformer, inductor and filter capacitor can be greatly reduced by operating at high frequencies. However, switching losses and iron losses increase, reducing circuit efficiency.

(a) Transformer

The basic design equation for a transformer with a square wave input is:

$$A \times N_1 = \frac{V_1}{4 \times f \times B_M}$$

where: A = iron cross-sectional area

   N_1 = number of primary turns

   f = frequency

   B_M = maximum flux density (avoiding saturation of the iron)

For a sinusoidal waveform this becomes:

$$A \times N_1 = \frac{V_1}{4.44 \times f \times B_M}$$

It can be seen from this equation that the size of the transformer is inversely dependent on frequency, therefore operation at high frequency will greatly reduce the transformer size.

Transformer efficiency, however, is reduced as frequency increases, as the eddy current and hysteresis losses both depend on frequency.

(b) Filter Inductor

For the continuous mode buck regulator:

$$L \geq \frac{R.(1-D)}{2.f}$$

Therefore the value of inductance needed decreases at higher frequencies.

In addition, for any given value of inductance, the size reduces at higher frequencies.

(c) Filter Capacitor

The output ripple voltage depends on the value of filter capacitance, the switching frequency, and the load.

For the buck regulator:

$$C > \frac{V_o}{\Delta V_o} \cdot \frac{(1-D)}{8.f^2.L}$$

$$> \frac{V_o}{\Delta V_o} \cdot \frac{1}{4.f.R}$$

Thus, by increasing the frequency, the capacitance can be reduced without increasing the ripple voltage.

(d) Switching Transistors

Typical transistor switching waveforms are shown below, including the power dissipated in the device ($P = v_T.i$).

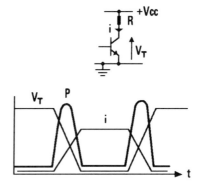

Voltage, current and power waveforms of a transistor during switching

Clearly, the semiconductor switching losses will increase with frequency. Switching devices are continually improving, particularly with the use of MOSFETs instead of bipolar transistors, so this limitation on switching frequency is much less important than it was a few years ago.

(e) Electromagnetic Interference

In addition, problems due to EMI increase as the switching frequency increases.

The reduced efficiency of high frequency SMPS creates problems of cooling the switching devices. This may require the use of larger heatsinks and/or the inclusion of fan cooling, which will add to the size and weight of the power supply.

Most power supplies are designed to operate with switching frequencies between 40 kHz and 200 kHz, although some designers are experimenting into the Megahertz range.

DESIGN EXAMPLE

Specification

Input voltage	300 volts $\pm 10\%$
Output voltage	5 volts
Output voltage ripple	< 50 mV peak-to-peak
Switching frequency	100 kHz
Output current	3 A to 30 A

Topology

Maximum output power = 150 watts

Therefore, choose a *Forward Converter*

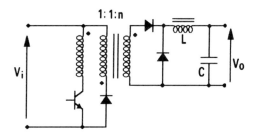

Forward converter (with tertiary winding)

Transformer Turns Ratio

For a forward converter:

$$V_o = n.D.V_i$$

But: $D < 0.5$

Therefore, taking $D_{max} = 0.45$, and allowing for a 0.8 volt drop across the secondary diode and the inductor resistance, the worst case is:

$$5.8 = n \times 0.45 \times (300 - 10\%)$$

Which gives $n = 0.048$

Inductor

For continuous mode operation:

$$L > \tfrac{1}{2}.R_{max}.T.(1-D_{min})$$

$$R_{max} = \frac{5 \ volts}{3 \ amperes} = 1.67 \ ohms$$

$$D_{min} = \frac{V_o}{n.V_{i(max)}} = 0.37$$

Therefore, $L > 5.3 \ \mu H$

Choose $L = 10 \ \mu H$

Capacitor

The output ripple voltage is given by:

$$\Delta V_o = \frac{V_o.T^2.(1-D)}{8.L.C}$$

Therefore:

$$C > \frac{V_o}{\Delta V_o} \times \frac{T^2}{8.L} \times (1-D_{min})$$

Gives $C > 79 \ \mu F$

Ripple Current

During t_{off}, $V_L = -5 \ volts$

$$V_L = L.\frac{di}{dt} = L \times \frac{\Delta I}{t_{off}}$$

Gives:

$$\Delta I = 3.15 \quad amperes$$

Capacitor ESR

$$\Delta V_o = ESR \times \Delta I$$

Gives:

$$ESR < 15.9 \ m\Omega$$

Transistor Ratings

$$Voltage = 2 \times V_{i(max)}$$
$$= 660 \ volts$$

The transistor should be rated to take into account any voltage spikes which may be present, therefore a voltage rating of 1000 volts would be suitable.

$$Current = n.\left(I_o + \frac{\Delta I}{2}\right)$$

$$= 1.58 \ amperes$$

In rating the transistor, allowance should also be made for the transformer magnetising current.

1.7 FUTURE CIRCUIT TRENDS

CUK CONVERTERS

Recently, much interest has been excited by the excellent characteristics of the Cuk converters, developed by Professor Slobodan Cuk of California Institute of Technology.

The basic Cuk regulator is derived from the boost and the buck regulators, combining the characteristic low input current ripple of the boost regulator with the low output current ripple of the buck regulator.

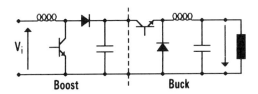

A combined boost-buck regulator

The boost-buck circuit above can be simplified to the Cuk regulator below.

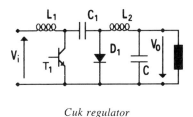

Cuk regulator

During the transistor on time, inductor current I_{L1} builds up according to

$$\frac{dI_{L1}}{dt} = \frac{V_i}{L_1}$$

Simultaneously, capacitor C_1 discharges round the loop C_1, T_1, C_2 in parallel with load R, L_2, charging C_2 in the opposite polarity to V_i .

When T_1 is switched off, inductor current I_{L1} flows through C_1 and the diode, recharging C_1 . Simultaneously, inductor L_2 maintains current round the loop L_2 , diode D, C_2 in parallel with load R.

This circuit has a very low output current ripple due to the presence of L_2 in the output circuit, similar to the buck regulator. It also has a very low ripple current in the input circuit due to the presence of L_1 , similar to the boost regulator. A further advantage is that only one transistor is required, with no base drive coupling problems as in the bridge and half-bridge circuits.

The d.c. gain of a Cuk regulator is (neglecting losses):

$$\frac{V_o}{V_i} = \frac{D}{1 - D}$$

where V_o is of the opposite polarity to V_i. This is the same as the basic flyback (buck-boost) regulator, but without the high input and output current ripple of the flyback.

A disadvantage of the Cuk regulator is that the series capacitor is the main energy storage/transfer component, instead of an inductor as in most other circuits. This capacitor, therefore, has to be relatively large, and capable of handling high r.m.s. currents, so a low ESR is essential.

By coupling the input and output inductors on a single core, a more efficient and compact design can be achieved for the same power throughput. In addition, the input and output ripple currents are considerably reduced.

The Cuk converter with isolation is shown below.

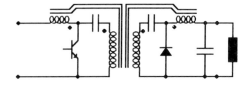

Cuk converter with the isolation transformer integrated
with the input and output inductors

The circuit operates in a very similar fashion to the basic Cuk regulator, with the coupling capacitor split up either side of the transformer. The presence of a capacitor in series with the primary (and secondary) windings prevents any d.c. components in the transformer, which might lead to the transformer saturating.

If the transformer is magnetically coupled to the two inductors, it is theoretically possible to tune the input and output current ripples to zero. In practice, very low values can be readily achieved. This very attractive feature removes the necessity of a large value, low ESR output capacitor, and greatly reduces the EMI problems on the input and output leads.

RESONANT CONVERTERS

The advantages of operating with a high switching frequency include:

- reduced transformer size,

- reduced inductor size,

- reduced capacitor size,

- improved transient response.

However, a high switching frequency suffers from two main drawbacks:

- increased switching losses,

- increased EMI problems.

The use of **Quasi-Resonant Converters** reduces the effects of both these problems.

Transistor Switching

When a transistor is switched from the off-state to the fully conducting state, there is a period when the current through the device has risen before the voltage across the device has dropped to zero, as shown below. The power dissipated in the transistor during this interval ($power = V.i$) can be appreciable, and at high switching frequencies can lead to the transistor overheating.

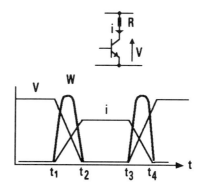

Voltage, current and power waveforms of a transistor during switching

The exact shape of these curves depends on the type of transistors used (e.g.

MOSFETs or bipolar transistors).

In quasi-resonant power supplies, the main switching waveforms are designed to be sinusoidal rather than square, with transistor switching taking place at either natural current zeros (zero-current switched converters), or natural voltage zeros (zero-voltage switched converters).

Zero-Current Switched (ZCS) Converters

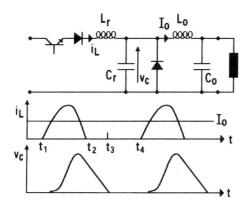

Half-wave, zero-current switched buck regulator

The circuit shown above is a conventional buck regulator, with a zero-current switch added.

The circuit operates as follows:

t_1 The transistor is switched on. Current i_L cannot increase instantaneously due to the inductor L_r, therefore the load current I_o continues to flywheel through the diode. Voltage v_C is therefore held at zero, so I_L builds up linearly:

$$V_{in} = L.\frac{di_L}{dt}$$

When $i_L = I_o$, current ceases to flywheel through the diode, and excess current $(i_L - I_o)$ charges capacitor C_r.

Both i_L and v_C vary sinusoidally at the resonant frequency until t_2, with v_C peaking when $i_L = I_o$.

t_2 $i_L = 0$, therefore I_o is supplied from C_r, discharging C_r linearly. The transistor will cease conducting at t_2. The transistor base drive should be removed between t_2 and t_3, to prevent it conducting again at t_3.

t_3 $v_C = 0$, therefore I_o flows through the flywheel diode.

t_4 The transistor is switched on again, and the cycle repeated.

It can be seen from the above that both switch-on and switch-off occur at natural current zeros, greatly reducing the transistor switching losses. In addition, as there are no very fast edges, EMI is also considerably reduced.

The diode in series with the transistor prevents the current through the inductor reversing, so the result is ½-wave zero-current switching. If the series diode is moved so that it is in anti-parallel with the switching transistor, as shown below, current through the inductor can reverse, giving full wave zero-current switching.

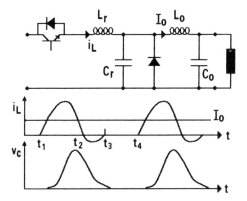

Full wave, zero current switched buck regulator

An advantage of full wave ZCS converters over ½-wave ZCS converters is that the output voltage in the former is not dependent on loading, giving very good load regulation. The d.c. gain becomes:

$$\frac{V_o}{V_i} = \frac{f_s}{f_n}$$

where:
f_s = switching frequency
f_n = resonant frequency = $\dfrac{1}{2\pi\sqrt{L_r C_r}}$

Thus, regulation is achieved by *frequency modulation,* rather than by pulse width modulation at constant switching frequency as with more conventional SMPS.

Zero-Voltage Switched (ZVS) Converters

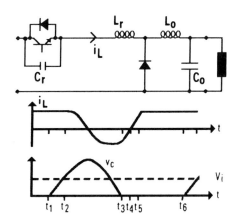

Zero voltage switched buck regulator

The circuit above operates as follows, starting with the transistor in the "on" state:

t_1 The transistor is switched off. The voltage across the capacitor v_c cannot change instantaneously, therefore the voltage across the transistor during turn-off is zero.
v_c will increase linearly, due to the approximately constant load current I_o.

t_2 $v_c = V_i$, therefore the flywheel diode is no longer reverse biased and will start to conduct load current. i_L will therefore start to fall, with C_r and L_r resonating.

t_3 $v_c = 0$, and cannot go negative due to the anti-parallel diode. Current i_L increases linearly.

The base/gate drive to the transistor should be applied during this period (as the anti-parallel diode is conducting, it will not conduct until t_4).

t_4 The transistor starts to conduct again.

t_5 $i_L = I_o$, therefore the flywheel diode is reverse biased.

t_6 The transistor is switched off again, and the cycle repeated.

It can be seen from the above that both transistor turn-on and turn-off occur at natural voltage zeros, therefore switching losses are greatly reduced.

Output voltage control is achieved by controlling the period (t_6-t_5), ie. by frequency modulation.

Resonant Converters With Isolation

There are a large number of resonant switching circuits which include an isolation transformer, an example of which is shown below:

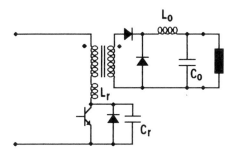

Half-wave, zero voltage, switched forward converter

This is a conventional forward converter, with ½ wave, ZVS included. Operation is very similar to the ZVS buck regulator described above.

Summary

Many different switching circuits have been developed using resonant techniques. ZCS and ZVS converters achieve the desired aim of greatly reducing both switching losses and EMI problems, at the cost of increased complexity. A further disadvantage of most resonant power supplies is that energy is transferred for a comparatively small percentage of the cycle compared to a conventional full wave converter (such as the bridge converter). Therefore resonant power supplies tend to have higher peak currents (hence higher i^2R losses) and/or peak voltages.

ZVS converters have an advantage over ZCS converters at high frequencies, due to the internal capacitance of the switching transistor. With ZCS, the energy stored in this capacitance is dissipated in the transistor at switch-on, adding to the switching losses. This problem is eliminated with ZVS.

A further advantage of ZVS at high frequencies is that it is possible to use the transistor capacitance as the resonant capacitor, avoiding the use of a discrete device. In addition, circuits with an isolation transformer can use the leakage inductance as the resonant inductor. A very compact, efficient design can thus be achieved.

ZCS converters are typically used up to about 1 MHz; however, ZVS converters have been built which operate in excess of 10 MHz.

SECTION 2

COMPONENTS FOR SMPS

2.1 MAGNETICS: DEFINITIONS AND EQUATIONS

PARAMETER	SYMBOL	UNIT
Magnetic flux density	B	$tesla$, (T)
Magnetic field intensity	H	$ampere\ m^{-1}$, (Am^{-1})
Permeability	$\mu = \dfrac{B}{H}$	$henry\ m^{-1}$, (Hm^{-1})
Permeability (free space)	μ_o	$henry\ m^{-1}$, (Hm^{-1})
Permeability (relative)	μ_r	$\mu = \mu_o \mu_r$
Effective magnetic area of core	A_c	m^2
Mean magnetic path length	l_e	m
Air gap length	l_g	m
Magnetic flux	Φ	$weber$, (Wb)
Magnetic potential	$m.m.f.$	$ampere$, (A)
Inductance	L	$henry$, (H)
Inductance index	A_L	$nH\ turn^{-2}$
Wire csa	A_x	m^2
Number of turns	N	
Mean length of turn	l_t	m
Current density	J	$ampere\ m^{-2}$, (Am^{-2})
Resistivity	ρ	$ohm -m$, $(\Omega - m)$
Energy	W	$joule$, (J)

REVIEW OF BASIC ELECTROMAGNETICS

Wound components, transformers and inductors (chokes) for use in SMPS are designed using sets of rules developed by engineers and characteristics supplied by manufacturers of magnetic cores and of wires. Because of this, it is not usual for designers to make complex theoretical electromagnetic calculations.

The techniques described in this book follow this methodology, and a brief description only is given of the electromagnetic concepts associated with transformer and inductor operation.

Magnetic Fields

Single Wire

The simplest form of a magnetic field is that around a straight wire. The diagram below shows the end view of the magnetic field near a straight-line conductor, where the field is set up by N amperes. (The older term of "ampere-turns" was probably more help in understanding magnetic fields: N ampere-turns can be set up by *one* wire carrying N amperes, N wires, each carrying 1 ampere or any combination of number of wires and number of amperes with the product N. The reason for the older term of ampere-turns should now be apparent.)

The field is termed *solenoidal* in that lines of magnetic flux are considered to exist co-axially around the wire. As with all fields, equipotentials meet flux lines at right angles. Thus, magnetic field equipotential surfaces may be considered as emanating radially from the wire. Magnetic flux lines are *always closed loops and are normal to the magnetic field equipotentials*. Inductance is defined in terms of the magnetic flux linked for each ampere flowing in the wire, as so even straight wires have an associated inductance.

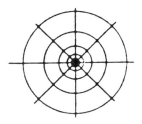

Magnetic field around a straight-line conductor

Magnetic circuits

For any magnetic path, Ampere's Law states that the sum of the different magnetomotive force (m.m.f.) drops around the path is equal to the total magnetic potential

$$m.m.f. = \int H.dl = Hl = NI \ ampere$$

The total magnetic field is independent of the path taken around the conductor. In the above figure magnetic field intensity, H, is inversely proportional to the distance, r, from the centre of the conductor,

$$H = NI/l, \qquad l = 2\pi r$$

$$H = \frac{NI}{2\pi r} \qquad ampere \ / \ metre$$

B-H relationship

The flux density, B, is proportional to the magnetic field intensity and a property of the material(s) through which the flux lines pass, viz, the permeability of the material(s):

$$B = \mu H, \ tesla$$

where μ is the permeability of the magnetic medium. Usually, μ is expressed relative to the permeability of free space, μ_o, i.e.

$$\mu = \mu_o \mu_r$$

where μ_r is the relative permeability of the material.

In the present work, the following assumptions are made and are taken as valid in the region of the B-H curve of interest:

    (i)   the flux density, B, is uniform
    (ii)   the field intensity, H, is uniform
    (iii) the permeability, μ, is constant, ie. linear B-H characteristic

2.2 WOUND COMPONENTS

Electromagnetic Induction

For a time-changing magnetic field, there is a voltage inducted in any wire within the field. The magnitude of this voltage is given by Faraday's Law of Induction

$$E = N\frac{d\phi}{dt} = NA_c \frac{dB}{dt} \ volt$$

The Inductor: Energy storage within a magnetic field

Energy storage within a magnetic field is particularly important when considering inductors. Power may be derived from the product of voltage and current: energy is then the time integral of power.

$$Edt = NA_c \ dB$$

Also :

$$I = \frac{Hl}{N}$$

combining:

$$W = \int EI \ dt = \int NA_c \ dB.\frac{Hl}{N} = A_c l_e \int HdB$$

Note:

The energy stored per unit volume in the magnetic field is:

$$W/m^3 = \int HdB = 1/2 \ BH \quad joule/m^3$$

Now:

$$W = \frac{1}{2} BHA_c l_e = \frac{B^2 A_c l_e}{2\mu_o \mu_r} \ joule$$

$$W = \frac{1}{2} LI^2 \ joule \ (watt-second)$$

The capability of a component to store energy depends upon its inductance, which can be determined as follows.

$$1/2LI_2 = 1/2BHA_c l_e$$

and:

$$L = BHA_c l_e \left[\frac{N^2}{H^2 l_e^{\,2}} \right] = \frac{BN^2 A_c}{Hl_e}$$

$$L = \mu_o \mu_r N^2 \, \frac{A_c}{l_e} \quad henry$$

Air Gap

If a wire is wound onto a former of magnetic material (i.e. a material which has a significantly large value of μ_r, for example, a core of ferrite) and a small gap, l_g, is made in the core, the flux in the ferrite will be the same as the flux in the air gap. If the air gap is small, the flux will pass across the gap almost in straight lines and hence it may be said that the flux density in the gap will be the same as that in the ferrite. Because of the ratios of relative permeability between the gap (air = 1) and the core (ferrite = 3000, typically) the magnetic field strength in the gap will be much higher than in the ferrite,

$$H = \frac{B}{\mu}.$$

With typical values of permeability, to a first approximation, the ferrite can be ignored and all the m.m.f. can be considered to exist in the gap.

Thus:

$$H = \frac{NI}{l_g}$$

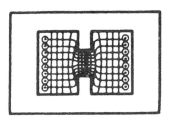

Winding with air gap

With all of the energy in the system stored in the gap, the energy density in the gap is easily calculated. The energy stored in the field outside the gap causes a few percent error in the calculation of stored energy and inductance but, in practice, a first-cut design is possible using this assumption (usually, the gap is much smaller in relation to the window area than shown in the figure).

The total energy is found by multiplying the energy density by the volume of the gap (centrepole area, A_c, times the gap length, l_g):

$$W = \frac{1}{2} BHA_c l_g$$

$$= \frac{1}{2} \mu_o H^2 A_c l_g$$

$$= \frac{1}{2} \mu_o N^2 I^2 \frac{A_c}{l_g} = \frac{1}{2} LI^2$$

i.e. again:

$$L = \mu_o N^2 \frac{A_c}{l_g}$$

For the present purposes, we will consider the *inductor* as an energy-storage device in which the energy is held within the magnetic field, that is the inductance values in the equivalent circuit represent energy stored in various physical regions of the device.

Notes:

The following points are important.

- the total energy stored in the gap equals one-half the product of the total flux, BA_c, and the total magnetic field, Hl_g.

- the energy *increases* if the gap is made smaller with the same ampere-turns in the winding. This is because the total field is the same, so the field intensity, H, must increase.

- the inductance of such an arrangement increases as the gap becomes smaller.

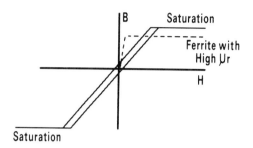

B-H characteristic for different permeabilities

B-H characteristic for different permeabilities

The B-H characteristic of a typical ferrite core with an air gap is shown in the figure. The dashed line is the characteristic of the ferrite alone, with high relative permeability and saturation at a relatively low flux density.

The solid line is the effective permeability, ie that value which would be obtained if the air gap were spread out over the entire length of the magnetic path through the core. For gaps typically encountered, the composite characteristic is dominated by the gap. Here the effective relative permeability of the composite is approximately the average of 1.0 in the gap and the much higher value of the ferrite over the entire path length.

The energy storage at any operating point is the area to the left of the curve from the origin up to the operating point, equal to the integral HdB, which simplifies to $\frac{BH}{2}$. For ferrites, the area to the left of the characteristic is very small, indicating that it is not possible to store significant energy.

This is ideal for a transformer, where energy storage is undesirable, but not for an inductor, whose main function is to store energy.

As we have seen, this storage capacity is enhanced by the introduction of a gap (air or other non-magnetic material) in series with the magnetic core. Some commercially-available powdered cores actually store energy in a series of microscopic "gaps" in the binder which holds the particles of core material together. Since this distributed gap cannot actually be measured, the manufacturer measures and specifies the *effective* permeability of the gap averaged around the length of the core.

As stated above, a gapless core does not store much energy, basically because its high permeability acts like a magnetic "short circuit." For such an arrangement, high flux densities can be achieved at very low field strengths: this property is ideal for a transformer.

Here, except for a relatively small magnetising current, the total ampere-turns in all windings add up to zero. The field caused by the ampere-turns of load current in the secondary windings is largely cancelled by the opposing field caused by primary winding current drawn from the power source. These two large opposing currents create a significant field between the two windings, as shown in the diagram. The energy stored in this field represents leakage inductance.

The Transformer

The ideal transformer has no winding resistance, has infinite primary inductance, and has perfect coupling such that all flux due to the a.c. excitation of the primary will link with the secondary winding. Under no load condition, an infinitely small current will flow. This current lags the applied voltage by 90° and is responsible for setting up the flux linking the primary winding with the secondary winding of the transformer. *This flux is constant for a constant applied voltage.* Since the flux is the same for the primary and the secondary, the induced voltages will be proportional to the primary turns N_1 and the secondary turns N_2, respectively. Therefore

$$\frac{V_2}{V_1} = \frac{N_2}{N_1}$$

For a loaded transformer, a secondary current flows:

$$I_1 N_1 = I_2 N_2$$

or:

$$\frac{I_1}{I_2} = \frac{N_2}{N_1}$$

But:

$$\frac{N_2}{N_1} = \frac{V_2}{V_1}$$

therefore:

$$\frac{Z_2}{Z_1} = \frac{\dfrac{V_2}{I_2}}{\dfrac{V_1}{I_1}} = \frac{I_1}{I_2}\frac{V_2}{V_1} = \frac{N_2^2}{N_1^2}$$

This is the concept of impedance transformation.

For a practical transformer, the primary inductance is finite, and the winding has finite resistance. A current will flow in the primary circuit even when there is no load on the secondary circuit; this current is called the *magnetising current*.

In general, if $v_1(t)$ is a time-varying voltage source, which, at a given instant, assumes the polarity shown in the diagram below , then the currents $i_1(t)$, $i_2(t)$, and $i_3(t)$ will assume the directions indicated and are increasing with time. The positive end of the winding is marked with a dot. *This is the dot notation.*

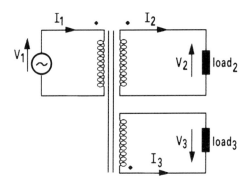

Transformer with dot notation

For a sinusoidal input voltage, the flux ϕ varies alternately:

$$\phi = \phi_{max} \sin\omega t$$

The instantaneous voltage induced in the primary is, according to Faraday's law,

$$e_1 = -\frac{d\phi}{dt} N_1$$

$$e_1 = -N_1 \phi_{max} \omega \cos\omega t$$

$$e_1 = -2\pi f N_1 \phi_{max} \cos\omega t$$

Therefore

$$E_{1max} = -2\pi f N_1 \phi_{max}$$

or, the r.m.s. value of $E_{1(max)} = E_1$:

$$E_1 = \frac{2\pi}{\sqrt{2}} f N_1 \phi_{max}$$

$$= -4.44 f N_1 \phi_{max}$$

For the general case, the applied voltage:

$$V_1 = K f n_1 \phi_{max}$$

where K ($=4.44$ for sinusoids, $=4$ for rectangular wave) is a constant.

The apparent power handled by the transformer is the sum of the primary volt-amps plus the secondary volt-amps.

For $N_1 = N_2 = N$, $I_1 = I_2 = I$,

$$P_t = V_1 I_1 + V_2 I_2$$

For a practical transformer, the efficiency is less than 100%, and:

$$P_i = \frac{P_o}{\eta}$$

where η is the transformer efficiency.

2.3 WOUND COMPONENT PRODUCTION

A major effort is required for both the design and production of magnetic parts in power supplies. It is important that consideration is given to the wound component requirements of any power supply as **EARLY** as possible in the design exercise in order that sufficient space be made available and that any difficulties with respect to weight are identified. As the switching frequency increases, the power transformer becomes the most difficult part to design. If the flux density is restricted to reduce self-heating, most of the common core materials - eg., ferrite, or powdered iron - are suitable up to, and above, 200 kHz.

Wire

The most commonly used material for windings is copper. When considering wound components, the resistivity of copper is important for two reasons;

(i) losses (and hence efficiency) of a wound component depends on, among other things, the resistance of the windings, and

(ii) regulation of the output voltage depends on any voltage drop in the resistance of the wound component winding.

Resistance of copper

The resistance of copper is dependent on temperature, viz at 20°C

$$\rho_{copper} = 1.724 \times 10^{-8} \Omega-m$$

At temperature, T°C:

$$\rho_{copper} = 1.724 \,(1 + 0.0042 \,(T-20)) \times 10^{-8} \Omega-m$$

Current density in the windings

The r.m.s. current density in a winding is limited because of the associated temperature rise.

An r.m.s. current density of 450 Acm^{-2} (*approximately* 3000 A in^{-2}) causes about 30°C temperature rise with natural convection cooling for a transformer or inductor whose core area product ($AP = W_a A_c$) is 1cm^2.

With larger cores, the current density for 30°C rise diminishes because the heat dissipating surface increases less rapidly than the heat producing volume.

While the high-frequency magnetic devices can be wound with standard magnetic wire, better copper utilisation is gained with Litz wire or multifilar magnetic wire or copper ribbon. These materials reduce skin effects, which increase with frequency and, above 100 kHz, may become problematical (difficulties may be experienced at even lower frequencies if the power is high). Grade 2 wire is generally used because it offers a reasonable compromise between a coating of sufficient thickness to avoid shorted turns while being thin enough to allow a practical winding pattern.

Winding

Difficulties are encountered in winding power transformers for higher frequencies not only in reducing the leakage inductance and capacitance but also in producing symmetrical and predictable voltage in the output windings. Several techniques are available, including interleaving, multifilar, twisted pairs and co-axial transmission lines.

Insulation requirements between windings:

Requirements are complex depending upon application and specification.

VDE 0806 and IEC 380, for office equipment, specify

Thickness: 3 layers, Mylar (0.016 cm with adhesive)

Creepage: 0.6 cm

Faraday shields: Copper foil (0.0076 cm with adhesive)

Frequency

Switched mode power supplies are available in the frequency range of a few 10's of kHz to 300 kHz with the lower end typical of the commercial market and the very highest operating frequency more likely to be met in the military field. Designs are now being studied in the 1-3 MHz range but these are not available as standard yet.

Two basic types of wound components must be considered viz, the transformer and the inductor - commonly termed the choke.

PRODUCTION OF TRANSFORMERS

Below is given information on typical transformers used in SMPS. It should be borne in mind that these data are not exhaustive and that many other materials and approaches to design can be adopted. However, effective and reliable designs can be produced if the methodology described in these sections is followed.

Material
For input transformers operating at mains frequency, soft magnetic steel is used. Typical at high-frequency are ferrites and dust cores with a range of compositions.

Operating Frequency
Low frequency - 50Hz.
High frequency - 20 - 100 kHz

Waveform
Usually square

Temperature
Operation is in elevated-temperature ambience, commercially around 60°C, military applications up to 100°C typical.

Current Density
450-950 Acm^{-2} with 775 Acm^{-2} as typical

Flux Density
$B_{saturation}$ = 0.42-0.45 T

Designs are restricted to approximately 60% of saturation level. This gives a safety factor to cope with any change in characteristic with increased temperature and to accommodate the magnetic conditions during overload or short circuit.

Thus, B_{usual} = 0.2-0.25 T at 20°C

Note $B_{saturation}$ reduces with increase in temperature

Core Shapes

E and I

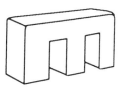

E core

These are normally used at mains frequency but may be encountered at high frequency from time to time.

Toroidal

Toroid Core

Size

Toroidal cores are available from small beads up to a maximum size of around 4 cm outside diameter.

Advantages

(i) Toroidal cores are generally used when heavy gauge wire is specified. This avoids the potential problem of shattering the fragile bobbins which are used with pot and RM cores.

(ii) Windings are brought out as flying leads to pins on suitable boards.

(iii) The main advantage of the toroidal shape is that it is available in different aspect ratios - deep with small diameter to shallow with large diameter - and can thus be fitted into unusual spaces. Cores can also be stacked for additional power.

(iv) With toroids good magnetic coupling can be achieved.

(v) Cores give minimum reluctance for a given core volume.

(vi) The parasitic capacitance of the wound component can be minimised.

Disadvantages

(i) Toroids are more difficult to wind than bobbins, especially for multi-layer construction.

(ii) Difficulty is experienced in insulating between layers.

(iii) Toroids are difficult to screen.

(iv) Mechanical fitting to boards can be difficult.

(v) Impregnation of the unit can give rise to high magnetising currents when ferrite is the core material. This is caused by the varnish or resin shrinking when curing and stressing the ferrite.

(vi) Achieving a suitable termination is more difficult than for the other core configurations.

RM- modular core

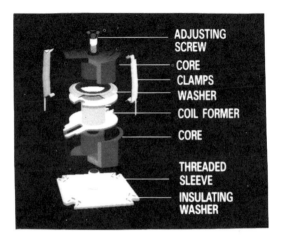

RM-modular core

These come with a bobbin for ease of winding. They are generally used because they can be mounted directly onto a p.c.b. It is usual to use double vacuum impregnation of the winding and bobbin configuration but potting of the components can lead to shrinkage and compression of the magnetic material. This, in turn, leads to an increase in magnetising current.

Windings may be made using high temperature materials to withstand the elevated temperatures of manufacture and operation.

Advantages

(i) Easy to wind, insulate and screen.

(ii) The winding on the bobbin can be impregnated separately from the core, thus avoiding the potential problem of core stressing.

(iii) Terminals are fixed to the bobbin and do not have to be provided by the designer.

Disadvantages

(i) Bobbins lack mechanical strength and can be damaged during winding, especially when using heavy gauge wire.

(ii) Bobbins lack mechanical strength to support heavy gauge wire.

(iii) The geometry of the wound component is restricted.

Pot cores

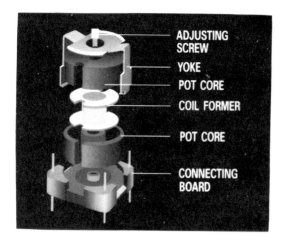

Pot core

As with the RM core, the pot core provides a convenient method of producing a transformer in that winding is very easy. There is more effective use of the window area because there is no layering effect as with toroids.

Pot cores provide slightly more iron than the corresponding RM core and have less associated leakage flux. Additionally, egress for winding wire is available at four points, 90° apart on the core, permitting steps of a quarter turn to be achieved. These factors mean that pot cores give maximum output for a given unit volume. Their major disadvantage is that the bobbins are straight so direct connection to a printed circuit board is not possible. Flying leads must be provided.

Wire

The universal wiring material is copper, but this is available in different configurations.

Shape

Round cross-section is normally used. Annealed tape wire of thickness 0.2mm is used when it is important to manufacture with more copper in the same window area. Litz wire, which has a stranded conductor, is used for higher frequencies. Unfortunately, it has a textile covering and so needs more room.

Skin depth

When the skin depth is smaller than the radius of the conductor, it is prudent to explore all the possible options for wire shape so that the most efficient use of the window area is achieved.

$$Skin \ depth \ (cm) = \left(\frac{6.61}{f^{0.5}}\right) K$$

where $K = 1$ for copper.

Coverings

Polyamide is used for specialist applications where high temperatures are expected (180°C and above) but it is tough and must be burnt off, then the wire cleaned with glass paper for termination.

Polyurethane is used for lower temperature applications (up to 150°C) and has the advantage of being self-fluxing which is particularly useful when using thin wires.

Terminations

As stated above, the RM core is provided with bobbin pins which connect direct to the printed circuit board. Standard resin-cored solder is adequate for all applications but, for high voltage devices, the resin must be washed off. When using fine wire, the end of the wire is taken back on itself so that there are three strands . These are twisted into a flex in which only one wire carries current, the other two contributing mechanical strength. Care must be exercised to avoid shorted turns and to solder spikes which may puncture the insulation.

Eyelets are soldered to wire terminations for low frequency (50 and 400 Hz) applications.

For toroidal cores, PTFE coated equipment wire is soldered to the winding wire, the joint secured mechanically and the PTFE coated wire brought out to the rest of the circuit.

Winding

Bobbins

Windings on bobbins have a layer of insulation material , normally about

0.012mm thick for fine wire and 0.025-0.05mm for heavier gauges, placed between layers of the winding.

For interwinding insulation, a suitable polyester tape may be used for insulation since these have high dielectric and mechanical strength.

Toroidal

Interlayer insulation may be with unimpregnated terylene tape.

Interwinding insulation uses the same material.

With high voltage devices, it is good practice to wind the primary on one side of the torus and the secondary on the other, although this reduces coupling efficiency. This technique is also used to minimise capacitive effects at high frequency.

PRODUCTION OF CHOKES

Material

Typical are gap ferrites and dust cores with a range of compositions.

Relative Permeability

Toroidal

μ_r values in the range 10 - 200 are possible depending on the choice of core material.

RM cores

When using RM cores, A_L values of up to 2000 $nH\ turn^{-2}$ are possible.

Operating Frequency

If low frequency (0 - 15 kHz) is used, choose a core material with high μ_r value, eg. 200.

If a high frequency circuit is to be considered , choose a material with lower μ_r; eg. for frequency of 100 kHz (and above), choose μ_r of 10. This is especially important in bridge circuits where the choke will be subjected to 2 x operation frequency, ie. normally 200 kHz.

The variation of μ_r with frequency is made to optimise the core design by attempting to match the copper and iron losses. Materials with a low μ value have high resistivity and thus lower core losses. In addition , such materials result in lower winding-to-winding and winding-to-core capacitance.

Waveform

Normally square, often with d.c. level.

Temperature

As for transformers.

Current Density

450-950 Acm^{-2} with 775 Acm^{-2} as typical. With chokes, it is important to take into account the IR voltage drop in the choke (due to output direct current), both for heating effects and for output voltage regulation. A typical value of IR drop is 0.1 - 0.15 V for a 12 V supply.

Flux Density

B_{max} of approximately 0.8 T at low frequency is usual for toroidal cores, with designs lying in the range 0.4 - 0.6 T.

At high frequency, $B_{saturation}$ = 0.6-0.7 T

B_{usual} = 0.3 - 0.4 T at 20°C for the RM core with a suitable gap. Design on the basis of all the m.m.f. being dropped in the air gap and none in the core, ie., the core merely provides a magnetic path for the flux.

With all designs take care to avoid core saturation when the design includes polarisation current.

Note $B_{saturation}$ reduces with increase in temperature.

Core Shapes

E and I

See transformers

Toroidal

These are widely used and have an inductance specified in nH per $turn^2$ (A_L value) to ease design.

With respect to *size, advantages and disadvantages* of toroids for choke manufacture, the same comments obtain as for transformers, described above.

RM- modular core

For choke manufacture, the centre limb has a gap. Otherwise, with respect to the general features of the RM core, same comments obtain as for their use in transformer manufacture, described above.

Pot cores

For calculation of inductance, manufacturers' tables should be used since the A_L value will vary with the size of gap in the core.

Otherwise, with respect to the general features of the pot core, same comments obtain as for their use in transformer manufacture, described above.

Wire

See transformers.

Winding

See transformers.

Testing

The following comments pertain to both transformers and chokes. Testing of chokes is done with polarisation current applied.

Voltage testing

Ionisation testing with up to 12kV between windings is possible.

Frequency

Up to 300 kHz is used.

Impedance measurements

a.c. and d.c. resistance of each winding

inductance of each winding

capacitance coupling between windings and between windings and the core

insulation resistance between windings (>10 MΩ)

Failure Modes

Manufacturing-based

shorted turns

incorrect number of turns

short circuits between windings or between winding and core

dirty terminals causing tracking (this can be avoided by careful cleaning of resin from terminals)

bad coatings on wires

Operation - based

overloading

circuit-generated voltage spikes

2.4   DESIGN EXAMPLES

TRANSFORMER

The design approach adopted is one of many possible, and it is not claimed that this is the *only*, or indeed, the *best* design methodology.

The example uses components from one specific manufacturer, but again, it should be emphasised that many other manufacturers produce components that are equally suitable and have characteristics that meet the specification to which the present designs have been developed.

Specification of the Transformer

The device specification will contain information on which the engineer will base his designs. This list is typical of the information which is supplied. If not supplied by the client, the designer should request it before starting out on the design. If some details are not available, the engineer should decide on suitable values, based on his experience and on discussions with the end-user.

Output VA

Although this may be derived from information given later, it is useful in classifying the design and selecting materials to know the secondary VA.

Frequency and Waveform

The primary switching frequency and the waveform details (eg, duty cycle) are required.

Primary Conditions

Input voltage, both steady state and transient.

Secondary Conditions

Information on the following should be made available:

(i) Number of windings

(ii) Number of taps per winding - e.g. centre tapped

(iii) Volts per tap per winding - ie. $V_{r.m.s.}$ per tap plus tolerances. Also, any restrictions on peak output volts should be noted., e.g. V_{opk} not to exceed 75V r.m.s. in worst case.

(iv) Regulation - Not normally a problem in practice, but requires consideration throughout design.

(v) Current per tap per winding - ie. $I_{r.m.s.}$ per tap plus steady state tolerances together with peak transient current.

For example,

$$I_{input} = 2A \ r.m.s. \ nominal$$
$$with \ 2.1A \ r.m.s. \ maximum, \ 1.9A \ minimum \ (steady \ state)$$
$$I_{output} = 3A \ (10s \ in \ 1h)$$

(vi) Type of load - e.g. full wave rectifier bridge or high impedance bridge

(vii) Impedance - As for primary, any limitations on impedance must be taken into account.

Mechanical Details

Often it is mainly mechanical restrictions which limit transformer design. The following information is required:

(i) Volume, ie. maximum height of component, maximum width of component, maximum length of component.

(ii) If the component is toroidal a minimum inner diameter is required

(iii) Method of fixing, eg. printed-circuit-board mounted or chassis mounted

(iv) Weight limitation

(v)   Terminations, eg. flying leads or former terminal pins

Environmental Details

The working environment must be described in order that component construction, materials etc., may be designated. The following details are used

(i)   Ambient temperature, ie. the temperature envelope including storage temperatures, eg. "Temp. = -40°C to +60°C"

(ii) Component finish, eg. dry or varnish impregnated or resin encapsulated

(iii) Vibration, frequency and amplitude

(iv) Screening, specify electromagnetic and electrostatic screening requirements

The following worked examples are illustrative of one particular approach, based largely on the methodologies specified by the SEI company.

It is emphasised that other approaches using other companies' materials are equally valid.

Design Specification:

$T_{ambient} = 90°C \ (maximum)$

Input

$140V \ (nominal) \ 150V \ (maximum), \ 122V \ (minimum)$

Output

$\pm \ 16V, 900mA \ (nominal), \ 1.2A \ (maximum)$: winding A

$\pm \ 22V, 100mA \ (nominal), \ 150mA \ (maximum)$: winding B

$+14V, 100mA \ (maximum)$: winding C

Primary Waveform

f = 100kHz, Duty = 50% as below

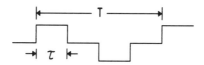

$T = 10\mu sec$ and $\tau = 2.5\mu sec$

The transformer may be redrawn to include diode drops:

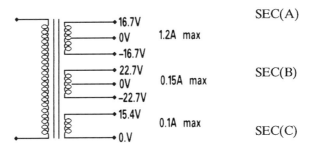

VA Calculation

The total secondary VA must be calculated.

$$Total = (16.7 \times 1.2) + (16.7 \times 1.2) \quad SEC(A)$$
$$+ (22.7 \times 0.15) + (22.7 \times 0.15) \quad SEC(B)$$
$$+ (15.4 \times 0.10) \quad SEC(C)$$

NB: The VA calculation uses peak current values.

$$= (40.08 + 6.81 + 1.54) = 48.43$$

Assuming efficiency = 85%,

$$Primary \quad VA = \frac{48.43}{0.85} = 56.97$$

$(I_p$ may be derived at this stage)

Core Selection

A core from the SEI range of RM ferrites is to be chosen - L2 grade.

L2 grade

Characteristics satisfy anticipated conditions:

1. T_c - Cure temperature losses at 100kHz not excessive

2. Mechanical form acceptable ie. printed circuit board mountable

Choose Suitable RM core from available range

(VA maximum = 70 approximately, given the above conditions - refer to RM selection chart).

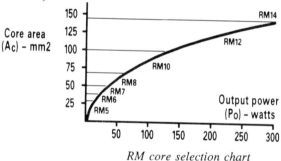

RM core selection chart

From the VA chart, an **RM8** core will be suitable, ie. RM core SEI M2 808/T/LT. The following data correspond to the selected core:

$$A_c = 64mm^2$$
$$B_{saturation} \ (20°C) = 520mT$$
$$B_{saturation} \ (100°C) = 340mT$$

Form Factor Calculation

Waveform exhibits 50% duty, i.e. T = 10μs, τ = 2.5μs

The following equations apply

$$form \ factor, ff = \frac{Vr.m.s.}{Vave.}$$

and

$$V_{r.m.s.} = V_{pk.} \sqrt{2\frac{\tau}{T}}$$

Hence,

$$V_{r.m.s.} = 140 \sqrt{\frac{(2 \times 2.5 \times 10^{-6})}{10 \times 10^{-6}}} = 98.99 volts$$

NB Working on **nominal** *peak values*

And,

$$Vave = \frac{2}{10 \times 10^{-6}} \int_{o}^{2.5 \times 10^{-6}} 140. \, dt$$

$$= \frac{1}{5 \times 10^{-6}} \times (140 \times 2.5 \times 10^{-6)} - 0$$

$$= 70 \; volts$$

$$Thus \, , ff = \frac{98.99}{70} = 1.414$$

Flux Density Selection

Estimated temperature rise at efficiency $\eta = 85\%$ is 40°C.

Hence, estimated operating temperature (worst case) = 130°C.

Thus, worst case $B_{saturation}$ = approximately 300 mT.

At this stage, knowing N_p, the peak flux density is derived. If $B_{pk.}$ is less than $B_{saturation}$, the design may proceed, but, *if $B_{saturation}$ is exceeded at any point of the flux excursion, unacceptable harmonics may be introduced.*

Turns Per Volt Calculation

$$Turns \; per \; volt = \frac{1}{4 \times ff \times B \times Ac \times f}$$

$$B_{pk.} = \frac{Vpk.}{4 \times ff \times Np \times Ac \times f}$$

Primary Turns Calculation

$$V_{prim} = 140V \ peak = 98.99V \ r.m.s. \ (99V)$$

Therefore using r.m.s. value,

$$N_p = 99 \times 0.184 = 18.22 \ turns$$

Rounding to nearest whole number,

$$N_p = 18 \ turns$$

Secondary Turns Calculation

Assuming 85% coupling,

$$SEC(A) \ V_A = +16.7V \quad 0V \quad -16.7V$$
$$N_{s(A)} = \frac{16.7 \times 0.184}{0.85} = 3.61 \ ie \ 4 \ turns$$

$$SEC(B) \ V_B = +22.7V \quad 0V \quad -22.7V$$
$$N_{s(B)} = \frac{22.7 \times 0.184}{0.85} = 4.91 \ ie \ 5 \ turns$$

$$SEC(C) \ V_C = 15.4V$$
$$N_{s(c)} = \frac{15.4 \times 0.184}{0.85} = 3.32 \ ie. \ 3 \ turns$$

[Note: The coupling factor will vary according to several factors - such as core geometry - and, as such, should be user-defined at this stage. For example, in above case, a coupling factor of 90% would yield $N_{s(A)} = 3.41$, ie. 3 turns or 3½ turns, which affects ratios and pin-allocations.]

Primary Winding Design

Assume the primary winding occupies 53% of total volume of the former. This is on the basis that the primary and secondary windings each should have half the available volume to allow heat dissipation. Since the primary carries the magnetising current as well as load current, a small extra allowance is made. Details of the dimensions of the former may be obtained from Manufacturer's catalogue.

Primary - assume 55% of total volume

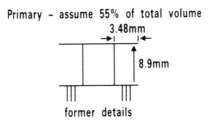

former details

Hence, primary depth should be around 1.84mm

Wire Selection/Build Calculations

$$Primary \; r.m.s. \; current, I_p = \frac{VA_{prim}}{V_{prim}} = \frac{56.97}{99} = 0.575A \; (575mA)$$

Standard wire sizes are available from a range of suppliers. From the sizes available, we can select the nearest wire size for $J = 450 \; Acm^{-2}$ (ie. $3000 \; Ain^{-2}$). This is wire with diameter = 0.40mm

(for a frequency of 100kHz, skin depth is around 0.40mm, so this choice of wire size means that there is maximum copper usage).

At this point, the constraints placed upon the designer by the manufacturing process must be considered. A deduction from the available depth, for factors such as insulation thickness on the wire, bulge etc., must be allowed. Depending on experience and the workforce available to the designer, this figure can vary over a range of values. We will choose 10% as a reasonable estimate.

Therefore, the depth which is available for primary conductor must take account of this bulge factor.

Details of 0.40mm Wire

Grade 2 insulation has a maximum diameter over insulation = 0.462mm. Therefore, the number of turns in 1 layer, including a deduction of 10% for slippage etc.:

$$\frac{(8.9-10\%)}{0.462} = 17.53 \; ie. \; 17 \; whole \; turns$$

Hence, for a primary of 18 turns (see above), this means 1 layer of 17 turns plus 1 layer of 1 turn will be required for the primary using 0.40mm Grade 2 copper wire.

Comments on First-Time-Through Design

Full occupation of the available volume will not take place using this wire. The choices are, therefore:

- to increase the wire size until 53% is achieved;

- to elect to use multistrand windings.

It is unwise to increase the wire size beyond 0.63mm because thicker wires present an excessive bulk to the former pins and place too much stress on the former. Neither is it desirable to have excessive bulk on pins due to too many wire strands.

It is good practice to choose the simplest solution which meets the electrical criteria.

For the present design, we will choose to use a single strand primary which fills the 53% volume as far as is practical.

Select 0.63mm Wire:

For 0.63mm wire, the maximum diameter (including insulation, Grade 2) is 0.706mm.

Hence,

$$Number \; of \; turns \; in \; 1 \; layer \; = \; \frac{(8.9 - 10\%)}{0.706} = 11.3, \; ie. \; 11 \; turns \; per \; layer$$

for 18 turns = 2, ie. the depth of the winding = 1.412mm.

This is within the limits set previously, and the winding will operate at a current density of approximately 190 Acm^{-2} (1200 Ain^{-2}; the ratio of cross-sectional areas of wire sizes: 0.40mm/0.63mm).

Hence,

a winding featuring single strand 0.63mm Grade 2 copper wire would be suitable.

Insulation

We must consider the insulation between primary layers. This is required owing to the existence of relatively high voltages (ie. as could exist between first and last primary turns).

Since the maximum temperature (ie. continuous temperature) is expected to be approximately 130°C, a class B system providing a minimum of 500 V isolation would be a distinct advantage.

Solution

There are several proprietary insulating tapes which meet this requirement (e.g., 3M No. 56 polyester tape which has a thickness of 0.06mm).

We can now define the primary winding:

Primary = 18 Turns in 2 layers, 0.63mm Grade 2 copper wire

1 layer 3M No. 56 polyester insulating tape over the first layer.

(Note 2nd layer of primary will be 'spread-out' to distribute turns evenly.)

Primary Cover

A Class B system is suitable. Also, to provide a 'bed' for the first secondary layer, a reasonable depth of insulation would be an advantage.

Hence,

2 layers No. 56 polyester tape should be used.

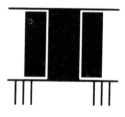

Primary winding

Total Depth

We must now check on the depth remaining for secondary windings.

Primary depth including an allowance for bulge

$$= (2 \times 0.706) \times 1.1 + (3 \times 0.06)mm = 1.733mm$$

This is less than the depth allocated above, i.e. 1.84 mm, so the winding should fit into the permissible volume.

$$(ie. \frac{1.733}{1.84} \times 100\%) = 94\%$$

Percentage occupation of volume allocated $= 94\%$, which is a good fill.

Secondary Windings

Available depth $= 3.48 - 1.733 = 2.107$mm.

In this transformer there are 3 secondary windings. The winding featuring the largest wire size (ie. largest current-handling requirement) is considered first.

[Normally, heavier gauges are positioned below lighter gauges to reduce stress on lighter gauges throughout winding. It is sometimes the case, however, that the opposite is true - for instance, where careful attention to coupling in a particular winding is required.]

16.7 V Winding

Again, the nearest standard wire will be chosen.

For $J = 450 \, Acm^{-2}$ (ie. 3000 Ain^{-2}), the nearest wire has a diameter $= 0.60$mm.

Including insulation, diameter $= 0.675$mm.

Hence,

$$Number \ of \ turns \ in \ 1 \ layer = \frac{(8.9-10\%)}{0.675} = approximately \ 12 \ turns$$

In this case, the layer would not be fully occupied.

We are approaching the upper limit for wire size on the former and the use of multistrand winding would be an advantage, since it may allow a better fill to be achieved. Although we will continue the design on the basis of a single wire, it is of interest to look briefly at the option of multistrand wire.

Multistrand option

Wire size suitable for complete layer-fill assuming double strands

$$= \frac{(8.9-10\%)}{16} = 0.506mm \ - insulation$$

Manufacturers produce multistrand wire in selected diameters. The nearest size is 0.425mm, (Grade 2: 0.489mm when insulation is considered).

$$Turns/layer = \frac{(8.9-10\%)}{0.489} = 16.6$$

Therefore

$$\text{Single layer depth} = 0.489\text{mm}$$

This would allow an alternative arrangement for the design of the winding. However, as stated above, we will continue the design on the basis of single strand wire.

Single-strand wire design

Design is

Secondary(A)

8 Turns, 0.6mm Grade 2 copper wire (1 layer) with centre tap

(with $J = 450\ Acm^{-2}$ (ie. 3000 Ain^{-2}), calculated as described previously)

Insulation

1 layer 3M No. 56 polyester (reasons for selection are as mentioned previously.)

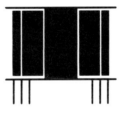

Secondary winding A

Total Depth

We must now keep a check on the depth remaining.

$$\text{Primary + cover + secondary(A) + cover}$$

$$1.733\text{mm} + (0.675) \times 1.1\text{mm} + 0.06\text{mm} = 2.535\text{mm}$$

Therefore,

Depth available for conductors in SEC(B) and (C) = 0.944mm

22.7 V Winding

Referring to wire tables, wire size for $J = 450 \, Acm^{-2}$ (ie. $3000 \, Ain^{-2}$) = 0.20mm

For 0.2mm wire, when insulation is taken into account, outside diameter is 0.245mm.

Hence,

$$Number \; of \; turns \; in \; 1 \; layer \; = \; \frac{(8.9-10\%)}{0.245} \; = \; approximately \; 33$$

Thus, we need only use one layer in which the turns are evenly distributed across layer,

ie. layer depth = 0.245mm

The design of the 22.7V winding is:
Secondary(B)

10 turns, 0.20mm Grade 2 copper wire (1 layer) with centre tap

Insulation

1 layer 3M No. 56 polyester (the reasons for selection are as mentioned previously).

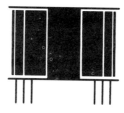

Secondary windings A and B

Total Depth

We must now keep a check on the depth remaining.

Primary, secondary(A) and insulation = 2.535mm

To this must be added:

Secondary(B) + cover,

which gives a total depth of
2.535 + (0.245) x 1.1mm + 0.06mm = 2.865mm

Therefore, remaining depth = 0.615mm

15.4 V Winding

Referring to wire tables, wire size for J = 450 Acm^{-2} (ie. 3000 Ain^{-2}) = 0.17mm.
Diameter over insulation = 0.211mm.

This winding will occupy a fraction of the available width. Hence, only 1 layer is required.

ie. layer depth = 0.211mm

Thus, the design is:

Secondary(C)

3 turns, 0.17mm Grade 2 copper wire

Insulation & Final Cover

**1 layer 3M No. 56 polyester tape (as previously discussed) plus,
as final cover, 1 layer 3M No. 79 woven-glass tape.**

Reasons for selection of woven - glass tape:

- meets temperature criteria

- meets electrical criteria

- material allows penetration of varnishing compounds

- material is tough

Material thickness = 0.06mm + 0.19mm

Primary + secondary(A) + secondary(B) + secondary(C) + insulation

= 2.865mm + (0.211) x 1.1mm + 0.25mm.

Total depth = 3.347mm

Note: Percentage occupation of former = 96%, which is reasonable and should provide a rugged component.

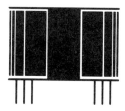

Secondary windings A, B and C

D.C. CHOKE : GAPPED RM FERRITE CORE

Choke Specification

INDUCTOR WITH DC BIAS

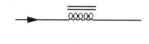

$f = 100\text{kHz}$ $\Delta I = 12.7\%$

Direct current $= 0.6$ A , Inductance, $L = 200\mu\text{H}$ (minimum),

Maximum voltage drop $= 0.1$ volt, Output voltage $= 5$ V r.m.s., $T_{ambient} = 40\,°C$

Core Choice

Select ferrite core with gap for energy storage - eg. RM core from SEI Ltd.

Material

We must select a material with characteristics which are suitable for operation at the design frequency (100kHz) and operating temperature (40°C ambient). SEI material L2 will be suitable and we will select it.

Size

Refer to data book curves LI^2 vs I^2R

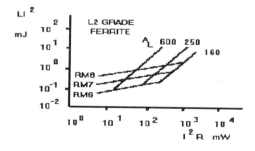

Core stored energy versus energy loss

$$LI^2 = 0.2 \times (0.6)^2 \text{ (L in mH, I in amps)}$$

$$= 0.072$$

I^2R loss which corresponds to the maximum allowable voltage drop of 0.1V at a current of 0.6A d.c. indicates a resistance R = 0.167 Ω.

Therefore, I^2R (maximum) $= (0.6)^2 \times 0.167 = 0.06$.

Therefore, from the curves, an RM6 core is suitable.

This is available with A_L = 100 or 160 nH $turn^{-2}$

Gap Choice

A large gap will support a larger energy storage (ie. LI^2 term) but the magnetising current is larger. Therefore, we will choose a smaller gap initially with A_L = 160 nH $turn^{-2}$.

Gap size = 0.20mm (from catalogue).

Turns Calculation

Using the A_L value of 160 nH $turn^{-2}$ we can calculate the number of turns required.

$$N = 10^3 \sqrt{\frac{L\,(mH)}{A_L}} = 10^3 \sqrt{\frac{0.2}{160}}$$

$$= 35.4, \text{ rounded to 36 turns.}$$

Flux Density Check

$$B = \mu H \qquad H = \frac{NI}{l_g}$$

Therefore $B = \mu \, H \, l_g = gap \; length$

Substituting gives $B = 135.7\ mT$

The maximum permissible flux density depends on the operating temperature, T_{op}. If the ambient temperature, $T_a = 40°C$, and a temperature rise of $40°$ is assumed,

$$T_{op} = 80°C$$

Saturation flux density at $80°C$, $B_{saturation} = 350$ mT approximately, from manufacturer's catalogue. Allowing a margin of safety of about two-thirds,

$$B_{op(max)} = 230mT$$

The ripple current is specified as 12.7% so the maximum total current will be 0.676 A.

$$B_{max} = 152.9\ mT$$

Hence, the flux density is below the design value and is thus acceptable.

Variation of Inductance with Current Estimate

Inductance varies with direct current, $I_{D.C.}$. From SEI data book curves RM6,

$$Inductance\ change = I_{dc}\ (mA).\ \sqrt{L\,(mH)} \times 10^{-6}$$

$$600.\sqrt{0.2} \times 10^{-6} = 2.683\ mT \quad \rightarrow \quad variation < 0.1\%$$

Thus, variation may be neglected.

Winding Design

For a direct current of 0.6 A, and a current density of $450\ Acm^{-2}$, the wire size = 0.425mm If we calculate skin effect at 100kHz it may be shown that only 0.40mm of the copper will be used. Therefore, there is nothing to be gained is using a larger wire; consequently we will select 0.40mm diameter copper wire.

In summary: Choose grade 2 copper enamelled giving an outside diameter of 0.462mm.

Former Details

RM6, 4 pin is suitable.

This has the following dimensions;

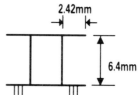

Layer width = 6.4mm
Layer height = 2.42mm

Allowances

Not all of the available volume can be used because of the practical limitations of winding. It is good practice to allow about a 20% reduction in the available height because of insulation and bulge.

Therefore height = 1.9mm

Similarly, it is good practice to allow about a 10% reduction in the available width because of slippage during winding.

Therefore width = 5.8mm

The number of turns which will fit a one layer of winding can now be determined.

$$Turns/layer \quad \frac{5.8}{0.462} = 12.5 \ ie. \ 12 \ turns/layer$$

Thus for the present winding, 36 turns will need 3 layers. This gives a winding height of

1.4mm

This is within the available height and is acceptable.

Insulation Requirements

The design is for a 5 V inductor and consequently electrical field strengths are not excessive (this is normally expressed in choke design as volts/layer). It follows that the insulation applied to the winding can be considered merely on the basis of providing mechanical strength for the building of the choke. Because of this, 1 layer polyester per layer (with thickness = 0.1mm) will suffice for insulation.

Therefore

Total height = 1.4 + (2 x 0.1) = 1.6mm

Insulating cover which gives mechanical strength = approximately 0.25mm.

Resistance Check

Copper wire of 0.04mm diameter has a resistance of 0.1372 Ω/m (at 20°C). This is temperature sensitive and allowance must be made for the change in resistance with temperature if the calculated resistance of the winding is close to the maximum specified on the grounds of voltage regulation.

Resistance of winding must be determined; for this we need mean length of turn (m.l.t.).

$$Turn\ length\ (1st\ layer) = circumference\ of\ former$$

$$= \pi D_f, \quad D_f = 7.45mm$$

$$= 23.4mm$$

Turn length (last layer)

$$= \pi(D_f + 0.462 + 0.100 + 0.462 + 0.100)$$

$$= 26.9mm$$

$$mean\ length\ per\ turn = \frac{26.9 + 23.4}{2} = 25.15mm$$

Therefore length of winding

$$= (25.15 \times 36)mm$$

$$= 905.4mm = 0.905m$$

Therefore

$$R_w = 0.905 \times 0.1372 = 0.124\Omega$$

Initial criterion was $R \leq 0.167\ \Omega$ so this design is within limits.

Final Specification

RM6 Gapped Ferrite Core

$A_L = 160 \ nH \ turn^{-2}$

Former = 4 pin

Winding 36 turns, 0.40mm Grade 2 copper wire

Insulation 1 layer polyester tape per layer of winding

Final cover 3 layers polyester

D.C. CHOKE TOROIDAL CORE

Specification

INDUCTOR WITH DC BIAS

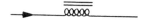

$$L = 0.15\text{mH}, \quad f = 75\text{kHz}, \quad \Delta I = 10\%, \quad I_{dc} = 2.5\text{A}$$

$$V = 15\text{V r.m.s.}, \quad \text{Maximum voltage drop} = 0.2 \text{ V r.m.s.}, \quad T_{ambient} = 30\,^\circ C$$

Core Choice

Toroidal core, iron alloy - eg. Genalex (SEI)

Important Points about Toroidal Cores

- heavier wire may be applied than on bobbin
- wide range of permeabilities available
- mounting of toroidal chokes is more difficult than bobbin type

From SEI Catalogue

Size of core depends on energy storage requirement and copper loss criteria.

The graph given by manufacturer is used :

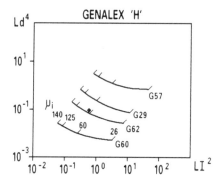

Energy storage versus copper loss

$Ll^2 = 0.15 \times (2.5)^2 = 0.975$

Ld^4 d = wire diameter in mm

If it is assumed that current density = $450\ Acm^{-2}$, d = 0.8mm.

(Choosing $450\ Acm^{-2}$ as opposed to a lower value, say $300\ Acm^{-2}$, limits wire size and increases copper losses. Whether this is acceptable or not will be seen later in design calculations.

$$\text{Hence,}\ Ld^4 = 0.15 \times (0.8)^4 = 0.061$$

From the graph, a core of G29 size upwards, with μ_i = 60 or under will be suitable.

Choosing a core of higher μ_i will yield more inductance in the same volume. d.c. saturation problems may result, however, and so, normally, a compromise has to be made. In this case, we choose G29 WH.

Details of G29 WH Core

$\mu_i = 50\pm5$, $l_m = 5.19cm = 0.0519m$, $A_{L(min)} = 24.8nH\ turn^{-2}$

Turns estimation

$$N = 10^3\sqrt{\frac{L(mH)}{A_L}} = 10^3\sqrt{\frac{0.15}{24.8}} = 77.8$$

say 78 turns

Check - Effect on μ

78 turns with 2.5A d.c. may prove to be excessive for this core:

$$For\ conditions\ specified,\ H = \frac{NI}{l_m} = \frac{78 \times 2.5}{0.0519} = 3757Am^{-1}$$

From manufacturer's chart shown below, at 3757 AM^{-1}, the WH grade will experience a drop in $\mu\Delta$ of approximately 15%. This may imply an unacceptable drop in inductance when d.c. current is applied.

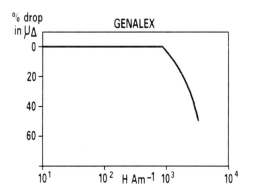

Variation of permeability with field strength for Genalex cores

If so, the VH grade may be considered or a larger core size used.

For example, choosing the VH grade would result in 106 turns. This increases copper loss (copper loss appears as I^2R heating) and also increases winding difficulty and cost. A larger core size improves the design but with the penalties of increased weight, volume and, normally, increased material cost.

For this example we will continue with the G29 WH design, ie:

78 turns, G29 WH core, 0.8mm wire.

Winding on the Core

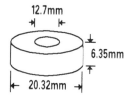

Core dimensions

outside diameter = 20.32mm
inside diameter = 12.70mm
height = 6.35mm

The number of turns per layer will vary since the inside diameter is decreasing per layer. Hence, it is difficult to be precise about winding resistances etc. - instead, assumptions as to the turns/layer must be made.

Layer 1

Inner circumference = πD = π x 12.7 = 39.9, say 40mm.

Allowing for slippage of conductors of 10% , usable circumference = 36mm.

For 0.8mm wire, Grade 2, the outside diameter = 0.885mm.

Therefore the number of turns in first layer = 40.6 (which is rounded to 40 turns).

Layer 2

Assume that first winding presents a uniform surface for layer 2 - ie. inside diameter is reduced by twice wire diameter uniformly.

Inside diameter now becomes : $\pi(12.7 - (2 \times 0.885))$ = π x 10.93

$$= 34.3mm - 10\% = 30.8 \text{ or } 31mm,$$

and so,

$$\text{turns in 2nd layer} = 31 \div 0.885 = 35 \text{ turns.}$$

Total number of turns so far = 75; this leaves 3 turns to complete the winding. These turns may be arranged to cover the entire core length, or be grouped together as required.

Resistance Check

Length of turn in 1st layer

$$= (2 \times 6.35) + \frac{2 \times (20.32 - 12.7)}{2}$$

$$= 20.32mm$$

Length of turn in 2nd layer

$$= 2 \times (6.35 + 2 \times 0.885) + \frac{2 \times (20.32 + 2 \times 0.885) - (12.7 - 2 \times 0.885))}{2}$$

$$= 27.4mm$$

Length of turn in 3rd layer

$$= 2 \times (6.35 + 3 \times 0.885) + \frac{2 \times ((20.32 + 3 \times 0.885) - (12.7 - 3 \times 0.885))}{2}$$

$$= 30.94mm$$

Hence, wire length $= (40 \times 20.32) + (35 \times 27.4) + (3 \times 30.94)mm$

$$= 1.864m$$

Resistance per metre $= 3.4 \times 10^{-2}\Omega$ *per metre* $(at\ 20°C)$ giving R $\simeq$ 63mΩ

Therefore at 2.5A d.c., voltage drop $= 0.1575$ volts, which is within specification.

$$I^2R = 0.39 \text{ watts}$$

Hence, the design meets the specifications. The following is worth pointing out; a larger core size could have been chosen and the addition of 2 or 3 extra turns on the above would increase inductance whilst probably remaining within specification in terms of I^2R .

Therefore final design is:

G29 WH Toroid

78 turns, 0.80mm Grade 2 copper wire, wound uniformly and continuously.

2.5 SEMICONDUCTOR DEVICES

Three types of semiconductor device dominate SMPS technology - the bipolar junction transistor (BJT), the metal oxide semiconductor field effect transistor (MOSFET) and the diode.

SEMICONDUCTOR DIODE

The most widely used type is the silicon diode. These devices are available in a wide range of current capabilities, ranging from tenths of an ampere to several hundred amperes or more, and are capable of operation at voltages as high as 1000 volts or more. Parallel and series arrangements of silicon diodes permit even further extension of current and voltage limits. Silicon diodes can be operated at ambient temperatures up to 200°C

Because of their high forward-to-reverse current ratios, silicon diodes can achieve rectification efficiencies greater than 99 per cent. They are very small and lightweight, and can be made highly resistant to shock and other severe environmental conditions. In addition, they have excellent life characteristics which are not affected by aging, moisture, or temperature.

Important Characteristics

Of the various diode characteristics for which data are given, for SMPS applications, the most important are thermal impedance, forward-voltage drop, reverse (leakage) current, and reverse recovery time. These four characteristics help determine the performance and environmental capabilities and limitations of diodes.

Thermal Impedance

Although silicon diodes can operate at high temperatures, the actual pellet of silicon which performs the rectification is quite small and has a very low thermal capacity. *During normal operation, the diode p-n junction dissipates approximately 1 watt of power for each ampere of forward current.* For safe operation the temperature of the junction should not rise above 200°C. For this reason, the silicon pellet is mounted between heavy copper parts in a symmetrical direct-soldered arrangement that results in uniform distribution of thermal stresses, minimum thermal fluctuations, and low thermal resistance.

The diagram shows a cross-sectional diagram of a typical silicon power diode.

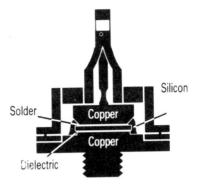

Cross-section of typical silicon diode

Manufacturers' specifications on silicon diode thermal resistance are expressed in °C per watt. When multiplied by the power dissipated by the diode, the rise in junction temperature above the case temperature is obtained.

Forward-Voltage Drop

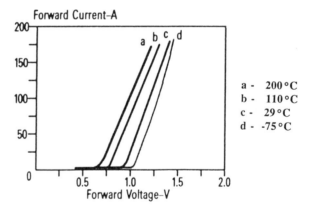

Forward characteristic of silicon diode

The major source of power loss in a silicon diode arises from the forward-conduction voltage drop, usually around 0.4 to 0.8 volt. The slope of the voltage-current characteristic at voltages above this threshold value represents the dynamic resistance of the diode.

The approximate expression for power losses P in a silicon diode, given by the following equation, can be used to explain how this type of operation is possible.

$$P_{(watts)} = (V_{dc}I_{dc}) = (I_{r.m.s.}{}^2 R_{dyn})$$

where the voltage V_{dc} is 0.4 to 0.9 volt depending upon the junction temperature, and R_{dyn} is the dynamic resistance of the diode over the current range considered.

Reverse Current

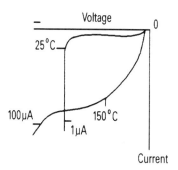

Reverse characteristic of silicon diode

Initially, as shown , reverse current increases slightly as the reverse voltage increases, but then tends to remain relatively constant, even though the reverse voltage is increased significantly. The figure also indicates that an increase in operating temperature causes a substantial increase in reverse current for a given reverse voltage. Reverse-blocking thermal runaway may occur because of this characteristic if the reverse dissipation becomes so large that, as the junction temperature rises, the losses increase faster than the rate of cooling.

If the reverse voltage is continuously increased, it eventually reaches a value (which varies for different types of silicon diodes) at which a very sharp increase in reverse current occurs. This voltage is called the *breakdown or avalanche (or Zener) voltage*. Although diodes can operate safely at the avalanche point, the device may be destroyed as a result of thermal runaway if the reverse voltage increases beyond this point or if the temperature rises sufficiently (eg., a rise in temperature from 25°C to 150°C increases the current by a factor of several hundred).

Reverse-Recovery Time

After a silicon diode has been operated under forward-bias conditions, some finite time interval (in the order of a few microseconds) must elapse before it can return to the reverse-bias condition. During this period, charge carriers in the device constitute a reverse current known as the reverse-recovery current.

The reverse-recovery time imposes an upper limit on the frequency at which a silicon diode may be used. Any attempt to operate the diode at frequencies above this limit results in a significant decrease in rectification efficiency and may also cause severe overheating and resultant destruction of the diode because of power losses during the recovery period.

Ratings

Ratings for silicon diodes are determined by the manufacturer on the basis of extensive testing. These ratings express the manufacturer's judgement of the maximum stress levels to which the diodes may be subjected without endangering the operating capability of the unit. The following list includes various factors for which silicon diodes must be rated: peak reverse voltage, forward current, operating and storage temperatures, amperes squared-seconds, and mounting torque.

Schottky Diodes

As operating frequencies of SMPS increase, the duty on the diodes has resulted in designers using diodes with better high-frequency performance. The Schottky diode is one of those devices found increasingly in SMPS circuits.

Structure

A Schottky diode is formed by placing a thin film of metal in direct contact with a semiconductor. The metal-semiconductor structure forms a low-resistance ohmic contact to semiconductor materials of all types. The operation of the device depends on quantum-mechanical effects which are beyond the scope of this book.

The arrangement has a rectifying $v - i$ characteristic very similar to that of a pn junction diode. The major difference is that at any given forward current, the voltage across the Schottky diode is smaller that that across a pn junction. The difference amounts to roughly 0.3 V, which means that Schottky devices are less lossy than conducting pn junction diodes. Thus, the Schottky diode may be preferable for use in some power applications.

In the reverse direction, the Schottky diode has a reverse leakage current that is larger than that of a comparable silicon pn junction diode. With present fabrication techniques, the breakdown voltage of a Schottky diode cannot reliably be made much larger than 100 volts.

Switching Characteristics

A Schottky diode turns on and off faster than a comparable pn junction diode. The basic reason is that Schottky diodes are majority carrier devices and have no stored minority carriers that must be injected into the device during turn-on and pulled out during turn-off.

The lack of any significant stored charge changes the shape of observed switching waveforms in important ways. During turn-off, there will be no reverse current associated with removal of stored charge. However, reverse current, associated with the growth of the depletion layer charge in reverse bias, will flow.

Schottky diodes have much less voltage overshoot during device turn-on than comparable *pn* junction diodes. The basic reason is that the ohmic resistance of the drift regions in a Schottky diode must be made much less than that of a *pn* junction diode in order to carry the same forward current because there is no excess carrier injection to short out high-resistivity drift regions. Some voltage overshoot associated with parasitic inductance will be observed if *di/dt* is large.

BIPOLAR JUNCTION TRANSISTORS (BJTs)

Both n-p-n and p-n-p bipolar junction transistors (BJT) can be used. The text is written on the basis of using an n-p-n BJT.

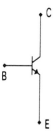

Symbol for an n-p-n BJT

The first two letters of the n-p-n designation indicates the polarities of the voltages applied to the emitter and the collector in normal operation. In an n-p-n transistor, the emitter is made negative with respect to both the collector and the base, and the collector is made positive with respect to both the emitter and the base.

The steady-state $v - i$ characteristics are as shown :

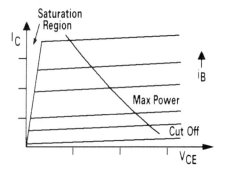

Characteristics of BJT

In SMPS applications, we are interested in the *switching* performance only of the active device.

The application of a sufficiently large base current (dependent on the collector current) results in the device being fully on. This requires that the control circuit provide a base current that is sufficiently large so that:

$$I_B > \frac{I_C}{h_{FE}}$$

where h_{FE} is the d.c. current gain of the device.

The on-state voltage, $V_{CE(sat)}$, of the power transistors is usually in the 1-2 V range, so that the conduction power loss in the BJT is quite small.

BJTs are current-controlled devices and base current must be supplied continuously to keep them in the on-state. The d.c. current gain h_{FE} is usually only 5-10 in high-power transistors and so these devices are sometimes connected in a Darlington, described later, to achieve a larger current gain. Some disadvantages accrue in this configuration including slightly higher overall $V_{CE(sat)}$ values and slower switching speeds.

METAL-OXIDE-SEMICONDUCTOR FIELD EFFECT TRANSISTORS (MOSFETs)

The circuit symbol of an n-channel MOSFET is as shown.

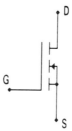

Circuit symbol of an n-channel MOSFET

The power MOS Field-Effect Transistor (MOSFET) is a device that evolved from MOS integrated circuit technology. The motivation for the development of these devices arose from the large base drive current required by power bipolar transistors and their limited switching speed capability.

Neglecting all the variation in structure and theory, there is one difference between a conventional bipolar transistor and a metal oxide field-effect transistor that greatly affects their performance in an operating environment. That is: the bipolar device is *current* driven, while the MOSFET is *voltage* driven. The gate of the MOSFET is electrically isolated from the source by a layer of oxide, giving a very high input resistance.

Characteristics of MOSFET

Again, we are interested in the switching performance of the transistor.

For the transistor in saturation, we can define:

$$g_m = \frac{1}{R_{DS(ON)}}$$

$$I_D = \frac{g_m}{2}(V_{GS} - V_T)$$

where $R_{DS(ON)}$ is the device on-resistance in the linear region. The voltage-current relationship of a conventional MOSFET is as shown:

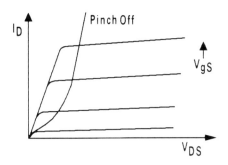

V-I relationship for a conventional MOSFET

Gate control is achieved by applying a positive voltage slightly in excess of the gate-to-source threshold voltage, $V_{GS(th)}$. Because of the relatively low drive voltage requirement and its high input impedance, the device is well suited for the control of high power directly from low-level logic circuits such as CMOS and TTL, provided adequate charge is supplied to charge up the input capacitance.

When a sufficiently large gate-source voltage is applied, the device turns fully on and approximates a closed switch. The MOSFET is off when the gate-source voltage is below the threshold value, $V_{GS(th)}$.

MOSFETs require the continuous application of a gate-source voltage of appropriate magnitude in order to be in the on-state. No gate current flows except during the transitions from on to off or vice versa when the gate capacitance is being charged or discharged. The switching times are very short, being in the range of a few tens of nanoseconds to a few hundred nanoseconds depending on the device type.

The on-state resistance $R_{DS(on)}$ of the MOSFET between the drain and source increases rapidly with the device reverse voltage rating. On a per unit area basis, the on-state resistance as a function of reverse voltage rating BV_{DSS} can be expressed as

$$R_{DS(on)} = k \ BV_{DSS}^{2.5to2.7}$$

where k is a constant that depends on the device geometry. Because of this, only devices with small voltage ratings are available that have low on-state resistance and hence small conduction losses.

However, because of their fast switching speed, the switching losses can be small. From a total power loss standpoint, 300-400-V MOSFETs compete with BJTs only if the switching frequency is in excess of 30-100 kHz. However, no definite statement can be made about the crossover frequency because it depends on the operating voltages, with low voltages favouring the MOSFET.

MOSFETs are available in voltage ratings in excess of 1000 V but with small current ratings and with up to 100 A at small voltage ratings. The maximum gate-source voltage is ±20 V, although MOSFETs that can be controlled by 5-V signals are becoming available.

MOSFETs are easily paralleled because their on-state resistance has a positive temperature coefficient. This causes the device conducting the higher current to heat up and thus forces it to equitably share its current with the other MOSFETs in parallel.

MOSFET Capacitances

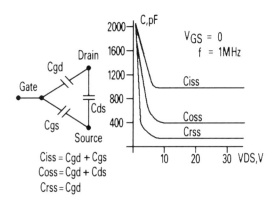

MOSFET capacitances

The physical structure of a MOSFET results in capacitance between the terminals. The metal oxide gate structure determines the capacitors from gate to drain (C_{gd}), and gate to source (C_{gs}). The pn junction formed during fabrication results in a junction capacitance from drain to source (C_{ds}).

These capacitances are characterised as input (C_{ISS}), output (C_{OSS}) and reverse transfer (C_{RSS}) capacitances on data sheets. The relationships between the interterminal capacitances and those given on data sheets, along with their variations as a function of drain voltage, are shown above. In driving a MOSFET, the input capacitance, C_{ISS}, is an important consideration. On a transient basis, gate current will flow to charge up this capacitance to gain gate voltage control. The generator, or driving impedance, and the input capacitance affect the switching capabilities of the MOSFET: the lower the generator resistance (R_{gen}), the faster the switching speed. Temperature variations have very little effect on the capacitor values.

Claims of infinite current gain for the power MOSFETs are valid only at low frequency. As the switching frequency rises, the gate drive circuit needs to be able to source and sink the pulse currents required to charge and discharge the high input capacitance of these devices. This is particularly frustrating, since it is at high frequency where the power MOSFET is particularly valuable due to its inherently high switching speed.

COMPARISON OF BJT AND MOSFET

Ideally, the transistor used as a switch will have the following characteristics:

high input impedance

low switching resistance

low on-resistance

high off-resistance

ability to withstand high overvoltages

rugged

Drive requirements

Bipolar power transistors (typically with current gains of 10 to 100) require a finite drive current but relatively low drive voltage. The base current often must be supplied from a power supply separate from the voltage on load. For applications requiring high current, considerable power may be dissipated in the base drive circuitry. Cascaded transistor arrangements are often used to reduce the base drive requirements.

In a MOSFET, the oxide layer that separates the gate from the channel region gives an input impedance of 10^9 to 10^{11} ohms as compared to 10^3 to 10^5 for the bipolar transistor. A MOSFET requires little current to maintain a constant voltage on its gate but current *is* required to change the gate voltage. Sufficient drive current must be sourced or sunk to change the gate voltage in the required time. A higher voltage (~ 10 volt) is also required to turn on fully a power MOSFET. The combination of high gate voltage and non-zero gate charge and discharge currents can often be met more easily than providing the drive to the base of a bipolar transistor, but these drive requirements must still be met.

Switching Time

The switching speed is directly related to delays within the structure and to capacitances which must be charged and discharged. The distance that carriers must travel from the emitter to the collector in a bipolar transistor is comparable to the source-to-drain distance in the MOSFET structure. Accordingly, any delays associated with carrier transit are similar.

Turning on a bipolar transistor requires that the capacitance associated with base-emitter junction be charged. The resulting charge distribution required to sustain current flow results in a "storage time" when the bipolar transistor is switched from " on " to "off". This storage time may be in the order of microseconds.

The switching time in a power MOSFET is determined by the gate capacitance and the current available from the drive circuitry.

Voltage Drop in the On State

Power MOSFETs have been portrayed as being capable of fast switching speeds, but having a relatively large forward drop in the on state. The same resistance value is present in the collector of the bipolar transistor and the drain of the MOSFET. However, the ability to bias the bipolar transistor into "quasisaturation" gives the bipolar transistor a significant advantage in voltage drop for a given current. This decreased voltage drop is not without its costs. The carriers present when the bipolar transistor is in quasisaturation result in a storage time. This storage time affects the device efficiency as a function of frequency. The power loss resulting from both switching and resistive losses means that the bipolar transistor has less total power loss at low frequencies while the power MOSFET is superior at high frequencies.

Device Off Resistance

Similar techniques are used in the manufacture of both power MOSFET and bipolar transistors. The silicon substrates used in device manufacture are capable of producing transistors with suitably low leakage currents at room temperature. However, at elevated temperatures, the leakage current increases, due to the thermal generation of carriers. This unwanted junction "leakage" sets one limit on the maximum transistor operating temperature. Bipolar transistors are typically rated at 150°C. Leakage currents do not affect the operation of power MOSFET devices as significantly. A power MOSFET with a rated junction operating temperature of 200°C has recently become available. There is no reason that this temperature rating cannot be achieved on all hermetic power MOSFET devices.

Transistor Breakdown Voltage

No electronics component is capable of withstanding an infinite voltage. The higher the voltage a device will withstand, the higher the voltage drop at a given current. It is most efficient to use a transistor with a voltage rating matched to the application. Investigation of the high voltage limits of bipolar and power MOSFETs shows that for the same material parameters in the collector and drain regions, there is a significant difference between the maximum voltage that the device can sustain under all operating conditions. The bipolar transistor collector region thickness and resistivity must both be increased if the sustaining voltage of the bipolar transistor is to equal the breakdown voltage of the power MOSFET. This change in the collector material reduces the voltage drop advantage that a bipolar transistor has over a power MOSFET.

At the time of writing, bipolar transistors are available with voltage ratings higher than MOSFETs. This greater availability is the result of at least four factors:

• Bipolar transistors have been under development for a longer period of time, so the techniques needed to obtain high voltages have been used more frequently.

• A market for high voltage bipolar transistor has been developed over the years.

• The large increase in MOSFET on-resistance with breakdown voltage results in poorer relative performance when compared to bipolar transistors as the breakdown voltage increases.

• The junctions used in the manufacture of MOSFETs are often shallower than those of a bipolar transistor. The surface geometry of a power MOSFET places a restriction on the junction depth if efficient surface utilisation is to be realised. High breakdown voltages are more easily obtained with greater junction depth.

With these constraints, present predictions are that power MOSFETs with ratings in excess of 1200V are unlikely to be commercially attractive.

Ruggedness

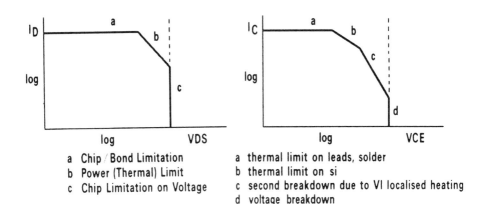

a Chip / Bond Limitation
b Power (Thermal) Limit
c Chip Limitation on Voltage

a thermal limit on leads, solder
b thermal limit on si
c second breakdown due to VI localised heating
d voltage breakdown

Safe operating areas of power MOSFET and bipolar devices

The allowable limits for simultaneous voltage and current levels during a transistor operating cycle, including switching, are defined by the Safe Operating Area characteristic. The manufacturer gives this for a range of operating conditions, from steady d.c. to short pulses.

When comparing these two device types in term of ruggedness, the power MOSFET transistor is superior. Bipolar transistors have a positive temperature coefficient with current while MOSFETs have a negative temperature coefficient. A bipolar transistor has substantially higher voltage blocking capabilities at low currents, but a power MOSFET is rugged at high currents. The negative temperature coefficient of the MOSFET reduces the current through localised regions, allowing the transistor to sustain a considerably higher current over the entire chip. The power dissipation characteristics of the package, whether it houses a bipolar or a MOSFET, is also of great importance. The difference in safe operating area between packaged bipolar and power MOSFETs is as shown.

With reference to the BJT SOAR characteristic shown, the constraints are provided in the four regions:

- The maximum allowable collector current. This is determined by thermal considerations for the mechanical parts of the device, ie. leads, solder, etc.

- The maximum allowable internal power dissipation in the silicon wafer. The resulting temperature rise will not cause chemical changes and damage.

- The second breakdown limit due to localised heating, with I_c and V_{ce} values both being high together.

- The upper operating voltage limit. This is normally the sustaining voltage, ie. the V_{ceo} maximum voltage. Operation above this voltage is only possible under restricted conditions.

The SOAR for switching is different and more extended than the SOAR for linear operation. In fact, during switching, the operating point has two stable low dissipation states and only transitionally traverses the zones of high dissipation, that is to say during a very short period of time compared to the thermal time constant of the transistor.

It is strongly recommended for all transistors whose voltage $V_{CEO(sus)}$ exceeds about 300V to operate within the low dissipation zone, particularly during the turn-off phase of the transistor.

The maximum junction temperature is the limiting temperature that the silicon device can tolerate. This temperature must never be exceeded, since the device could then be damaged or destroyed. Lower operating temperatures improve transistor reliability.

When the above recommendations are respected, even under worst case conditions, the switching transistors are a very reliable device.

Economics

A comparison of the process steps used in the manufacture of the two types of transistors reveals that both the process complexity and the number of critical process steps is greater for a MOSFET. For the same die size, MOSFETs now approach the costs of bipolar transistors and, in lower power devices, MOSFETs are becoming cheaper. The prevailing trend in costs is downwards for both types of device, as can be seen in the diagram.

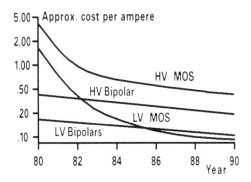

Cost trends of power semiconductors

Power MOSFETs were also found to exhibit superior safe-operating-area and output characteristics for paralleling when compared with the bipolar transistor. However, these merits of the power MOSFET were offset by a higher on-resistance per unit area and a much larger processing cost when compared to the bipolar transistor. For power MOSFET manufacturing the yield was much lower than for bipolar transistors of the same size. Consequently, high manufacturing costs of power MOSFETs restricted their application to high-frequency circuits (such as those used in switch-mode power supplies) and to low-voltage (<100 V) circuits where their on-resistance reached acceptable values.

With improvements made in process technology to obtain better yields and performance, the ratings of power MOSFETs have continued to grow over the last 10 years. At present, devices are available with breakdown voltages of up to 1000 V at current levels of about 1 A and current handling capability of over 20 A at breakdown voltages below 100 V.

SECTION 3

SMPS CONTROL & DRIVER CIRCUITS

3.1 DRIVE CIRCUITS

COMPARISON OF SWITCHING CHARACTERISTICS OF BJT AND MOSFET

For similarly rated devices, not only does the BJT require a greater current for switching, but the current must be maintained. In addition, the speed of both switch-on and, more noticeably, switch-off is slower for the BJT.

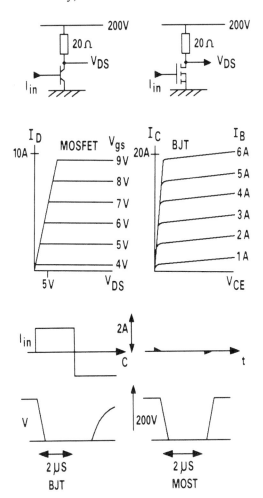

Comparison of switching characteristics

The unforced gain of a transistor in the conduction mode is equal to $\dfrac{I_c}{I_{B1}}$. During

turn-on, a large current pulse is applied; usually a force gain $\dfrac{I_c}{I_{B1}}$ of 2 to 5 during turn-on is sufficient. A point of diminishing return, where more overdrive does not significantly improve the turn-on time, is reached well before there is any danger of dissipating too much energy in the base.

Switching of BJTs

The central problem in high frequency converter design is how to obtain rapid state change in the switching of the bipolar power transistor. Two options are available to the circuit designer: first, to select a fast device, and second, to control the base drive to optimise the transition time and reduce the storage time.

Bipolar power transistors with potentially good switching performance are readily available. Finding a fast transistor is no real problem. Most of the design effort is directed towards controlling the base drive. The optimisation of the base drive is therefore the dominant factor in the design of a converter circuit. Its influence is often the determining factor for the reliability of the system.

Turn-On

The turn-on time is made up of the *delay time* and the *risetime*. The delay time is due to the finite time required to discharge the b-e junction depletion layer capacitance and for emission to commence, plus the transit time of carriers across the base. The risetime (10% - 90% of the final collector current) is the time required to neutralise the collector junction depletion layer and flood the base with carriers to sustain the required level of collector current.

Turn-on time can be minimised by increasing the initial value of base current; ie. some method of peaking should be provided to ensure the shortest possible turn-on time.

Turn-Off

The collector current does not respond immediately to the removal of base drive (current). Before I_c can fall, the minority carrier level in the base region must fall from the very high saturation level. The delay before this happens is the *storage time, t_s*. For fast turn-off it is necessary to extract this stored charge rapidly at the base by a reverse base current driven by a reverse b-e voltage. When this is completed, the b-e junction builds up a depletion layer, the reverse I_b decays to zero and the collector current falls to zero. The time, following the storage time, for the current to fall from 90% to 10% of the initial current is the *fall time t_f*. The total *turn-off time* is the sum of t_r and t_f, and is typically a few microseconds with $t_s > t_f$ usually.

N.B. Storage time and turn-off time are much reduced by having a minimum of stored base charge during conduction ie. that just required for saturation with $I_b = h_{FE} I_c$. This conflicts with requirements of fast turn-on and low V_{ce} during conduction.

Turn-Off Time

There are two methods to reduce the fall time of the collector current:

• During the conduction time the base current is held to the level required to maintain $V_{CE} \geq V_{BE}$ for any I_c, ie. the transistor is driven into quasisaturation (instead of hard saturation). In that case the quantity of minority carriers (stored charge) in its collector area is much reduced, storage time is improved, and a high negative base current (I_{B2}) can be applied, without the risk of a minority carrier tail.

• To avoid increased conduction losses due to the higher on-state voltage drop in quasi-saturation, the transistor is driven into hard saturation during the conduction time. At the beginning of turn-on the positive base current is removed and a light negative base current is applied to the transistor. This will lead to an increased storage time, but now minority carriers in the collector have time to recombine. At the end of the storage time, the near end of the recombination leads to an increase of the collector-emitter voltage drop. This increase is detected and then a very high negative base current is applied. This drive method gives very short fall times but the storage time t_s will be higher than with the quasi-saturated drive.

Common to both methods is the idea of applying high negative base current only when minority carriers in the collector area have been recombined.

Characteristics of Drive Circuits for Switching

To obtain current base drive, the following points should be respected:

(i) during the turn-on phase, the base current must have a high rate-of-rise and, if possible, an overshoot (2-3 μs long) which favours a fast transition;

(ii) during the conduction phase the base current must maintain the transistor in saturation or quasisaturation;

(iii) for the turn-off period, if the transistor is operated in hard saturation, $\dfrac{-di_B}{dt}$ of the base current waveform must be controlled and limited in value so that the collector and emitter currents are cancelled at the same time.

(iv) provide voltage isolation between the "signal" and "power" function.

(v) include protection against high dv/dt or surge voltages. The signal from the PWM control circuit (usually only several volts or milliamperes) must often be amplified to a level of several amperes required by the transistor. A base-current shaping circuit enables optimisation of the switching times of the transistor to be obtained.

Forward base drive is designed to put the transistor into saturated (or quasisaturated) operation mode. The base current is estimated by dividing the base-drive voltage by the dynamic resistance of the base circuit. With reverse voltage applied to the base, no base current flows in steady state because the b-e junction is reverse biased. However, during the transition in state associated with switching, reverse base current will flow.

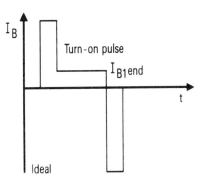

Idealised switching waveforms

In practice, the waveform shown is used:

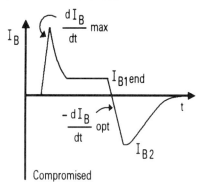

Practical switching waveform

The negative (reverse) drive is a compromise, and both the slope and peak value can be made to vary with the amount of stored charge to optimise the turn-off performance.

Turn-On Base Current Peaking

The simplified circuit illustrates the concept of base current peaking.

Less sophisticated methods are often employed, and in practice an approximation is made to the ideal waveforms, resulting in the waveform shown.

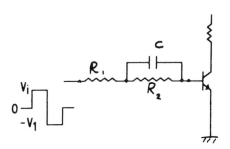

Circuit for base current peaking

The base peaking circuit consists of R_1, R_2, and C. When V_i goes positive, a fast voltage edge is applied to the base of the BJT via R_1 and C. The transistor turns on and, at the same time, C charges up. Eventually, when C is fully charged, the value of base current is set by R_1 and R_2.

Note that C charges up to a final value of:

$$V_C = V_1 \frac{R_2}{R_1 + R_2}$$

Once V_i returns to zero, (or goes negative), the base-emitter junction of the BJT becomes reverse-biased and C is forced to discharge through R_2.

In order for this scheme to be workable, certain conditions must be satisfied. The pulse width must be at least 5 times the time constant of the charging circuit, and the discharge time constant must be around 5 times the off-period of the input voltage.

Turn-off Circuit

The turn-on of the switching transistor Q_1, using R_1, R_2 and C_1 is as described above.

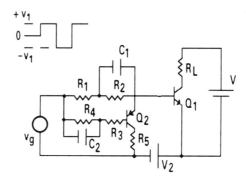

Turn-off circuit

As the input voltage goes negative, transistor Q_2 is driven "on" via the current peaking circuit of C_2 and R_3 initially, and then via R_3 and R_4. The reverse current through the base-emitter junction of Q_1 forces the transistor off.

An inductor is sometimes added in parallel with the resistor R_5. In this case R_5 is then made larger and used only as a damping resistor. The purpose of the inductor is to limit the negative slopes of the current i_{B2} so as to optimise the fall time of transistor Q_1 when this transistor is driven into hard saturation during its conduction time. The di/dt limiting effect of the inductor results in a controlled rate of extraction of the stored charge and allows the minority carrier in the collector of Q_1 time to recombine, thus avoiding a minority carrier tail.

Anti-Saturation Circuits

Anti-saturation circuits (clamping circuits) are used when maximum switching speed must be achieved for a transistor. The parameter most affected by operating in soft saturation compared to hard saturation is storage time. Substantial reductions in storage time can be made (as high as 25 to 1) by not fully saturating the transistor.

In the saturated switching mode, the circuits are simple and voltage levels are well-defined. Costs are reduced and the transistors have to dissipate less power, which means smaller heat sinks can be used.

The disadvantage of saturated switching is that storage time (which is proportional to forward base current) can be large and will limit the frequency of operation of the SMPS. Anti-saturation circuitry reduces storage time markedly.

The principle of anti-saturation circuits is illustrated by the circuit shown.

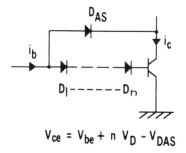

$$V_{ce} = V_{be} + n \, V_D - V_{DAS}$$

Anti-saturation circuit

When the transistor V_{CE} falls to within one diode forward voltage drop of nV_D, the diode D_{AS} conducts and limits V_{CE} to nV_D - V_{AS}. When the collector is clamped, current flows through the diode D_{AS} equal to:

$$I_{AS} = \beta i_B - \frac{V_{CC}}{R_L}$$

Therefore, heat must also be removed from D_{AS}. Since the collector clamp results in higher power dissipation in the transistor, nV_D should be as low as possible.

Darlington Configuration

The Darlington configuration is the only realistic way of operating very high current transistors, where the base drive requirement for a single device is extremely high e.g. I_c = 250 A, h_{FE} = 10, therefore required I_b = 25 A. Darlingtons are available in monolithic form (2 transistors on one chip), or can be built from two discrete transistors.

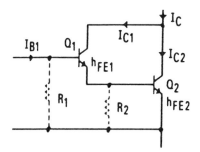

Darlington configuration

The (smaller) driver transistor Q_1 emitter current supplies the base of the 'main' transistor Q_2. The overall h_{FE} is greatly increased, reducing the base drive I_{b1}

requirement.

D.C. gain, h_{FE}

Let h_{FE1} $(= \dfrac{I_{c1}}{I_{b1}})$ be the d.c. gain for Q_1 and h_{FE2} $(= \dfrac{I_{c2}}{I_{b2}})$ be the Q_2 gain.

$$\text{Overall gain } h_{FE} = \frac{I_c}{I_{bl}}$$

$$h_{FE} = h_{FE1} + h_{FE2} + h_{FE1} \cdot h_{FE2}$$

Approximately, the Darlington h_{FE} is the product of the individual transistor h_{FE}'s.

BJTs, whether in single units or made as a Darlington configuration on a single chip [a monolithic Darlington (MD)], have significant storage time during the turn-off transition. Typical switching times are in the range of a few hundred nanoseconds to a few microseconds.

BJTs including MDs are available in voltage ratings up to 1400 V and current ratings of a few hundred amperes. In spite of a negative temperature coefficient of on-state resistance, modern BJTs fabricated with good quality control can be paralleled provided that care is taken in the circuit layout and that some extra current margin is provided that is, where theoretically four transistors in parallel would suffice based on equal current sharing, five may be used to tolerate a slight current imbalance.

Stabilising Resistors

These are shown dotted as R_1 and R_2, and stabilise the base potentials. Without R_1 and R_2 the total emitter current of Q_1 is injected into Q_2 base. Even with Q_1 cut off, its small leakage current into Q_2 base is amplified, giving a much higher overall leakage than for Q_2 alone. The overall h_{FE} is reduced very slightly by R_1 and R_2.

Collector-emitter saturation voltage

For transistor Q_2 alone, when saturated, $V_{ce2(sat)} < V_{be2(sat)}$ as its collector junction is forward biased. In the Darlington, with Q_1 driven hard and saturated,

$$V_{ce2} = V_{be2} + V_{ce1(sat)}$$

Hence, now $V_{ce2} > V_{be2}$, and Q_2 operates on the edge of saturation (quasisaturated), with Q_1 hard saturated; $V_{ce1(sat)}$ is low and V_{bc2} is the major component of V_{ce2}.

Thus, the penalty for greatly increased h_{FE} is a greatly increased power dissipation in Q_2 during conduction, due to higher V_{ce}. However, its heatsink can be cooled fairly easily and losses are better dissipated here than in the base drive circuit.

Switching Speeds

Turn-off time is increased considerably as the overall storage time is the sum of the individual storage times of Q_1 and Q_2. Turn-on time is also increased, though less so, as Q_1 collector current must rise appreciably before Q_2 base drive is sufficient for it to start its turn-on.

Speed-Up Diode

A diode connected as shown in the diagram, in anti-parallel with the base emitter junction of Q_1, enables a considerable reduction in the storage time of the Darlington to be achieved. In order to render this diode conducting, and consequently to evacuate the carrier stored in the base of Q_2, the base-emitter junction of Q_1 must first be completely cut off. A negative voltage can then only be applied when Q_1 is cut off. Despite the fact that switching always occurs in cascade, the diode speeds up the switching performance of the Darlington.

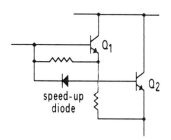

Speed-up diode

GATE DRIVE FOR POWER MOSFETs

The circuitry to drive MOSFETs in a switched mode initially appears simple and the applications of such devices to power supplies is an attractive proposition. Indeed, this *is* the case at low frequency, where the very high input resistance of the MOSFET means that very low current levels are required and the device is switched by the application of voltage alone.

However, the development of switched mode power supplies has involved a steady movement to higher switching frequencies because of the increased power densities possible. At such frequencies, the input capacitance of the MOSFET becomes the dominant feature of the input impedance, and driver circuits must be designed to have the current source and sink capability to charge and discharge, respectively, the effective gate-to-source capacitance.

Some integrated circuits, including those specifically designed for switched mode power supply control purposes can provide up to 100 mA of sink and source output capability, and when directly coupled a MOSFET can switch reasonably efficiently at 20 kHz. However, to switch efficiently at higher frequencies, several amperes of drive may be required.

Drive Requirements

A MOSFET responds instantaneously to changes in gate voltage and will begin to conduct when the threshold is reached (V_{GS} = 2 to 4 V) and be fully on when V_{GS} = 7 to 8 V. Gate waveforms will show a step at a point just above the threshold voltage which varies in duration depending on the amount of drive current available. The drive current determines both the rise and fall times for the drain current. The drive current can be estimated as follows:

$$I_m = C_{DG} \, dV/dt$$
$$I_G = C_{GS} \, dv/dt$$

where I_m is the current required by the Miller effect to charge the drain to gate capacitance at the rate it is desired to move the drain voltage, and I_G is the current required to charge the gate to source capacitance through the linear region (2 to 8 V).

As an example, if 30 ns switching times are desired at 300 V where

$$C_{DG} = 100 \text{ pF and } C_{GS} = 500 \text{ pF, then}$$

$$I_m = 100 \text{ pF x } 300 \text{ V/30 ns} = 1.0 \text{ A}$$

$$\text{I sub G} = 500 \text{ pF x } 6 \text{ V/30 ns} = 0.1 \text{ A}$$

This example shows that speed is directly tied to drive current and that C_{DG} will have the greatest effect on switching speed and that C_{GS} is important only in estimating turn-on and turn-off delays.

Drive Circuits

The drive requirements of the power MOSFET result in simple drive circuits, whether isolated or non-isolated.

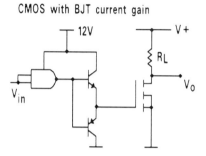

CMOS with BJT current gain

MOSFET drive circuit

The circuit shown derives a basic low-power switching waveform from CMOS logic and achieves the required drive current capability from a bipolar transistor pair. If voltage isolation is required this can be achieved with a transformer.

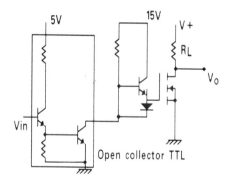

Improved MOSFET drive circuit

There are applications where higher drive capability must be provided. The circuit shown is an improvement on the CMOS drive circuit above. Here, current gain is again provided by bipolar transistors with the final drive stage being fed from a higher voltage supply, with corresponding increased drive capability, especially where high-power MOSFETs are used.

SNUBBERS

High frequency power conversion circuits subject power transistors to high instantaneous power dissipation during switching, both turn-on and turn-off. Additionally, as semiconductor manufacturers continually make their devices faster, the safe areas of operation must be carefully considered to prevent device failure. Consequently the circuit designers must pay particular attention to the turn-on and turn-off load lines to ascertain that they lie within the appropriate safe regions of the operation area. If the SOAR limits are being exceeded, load line shaping by the use of turn-on and turn-off snubbers must be employed.

Snubbers are passive switching aid circuits composed of diodes, resistors and impedances (capacitors or inductors) which reduce the severity of electrical effects on the transistor while switching. Of particular interest are;

• the rate of increase of voltage dV/dt while the transistor is being turned off.

• the rate of growth of the current dI/dt while transistor is being turned on.

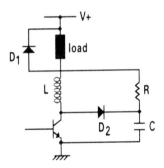

Unified snubber circuit

Shown is a unified snubber circuit: this has two principal functions, viz:

• it reduces the rate of growth dV/dt of the voltage on the transistor for the turn-off time, and consequently allows the use of a transistor having a lesser safe operating area.

• it allows the considerable reduction of losses in the transistor. It could be said that, approximately, the switching losses are transferred from the transistor to the resistor R.

For other types of application, other advantages become apparent; parallel connection of transistors is made easier, reduction of EMI/RFI, etc.

The reduction of losses in the transistor gives the following advantages: better reliability and faster switching times, both of which benefit from a fall in junction temperature.

In most switched mode power supply circuits the switching transistor load lines are inductive during turn-on because the short circuit created by output diode reverse recovery times is isolated by leakage inductance in the transformer. The inductance effectively snubs the turn-on load lines so that the diode recovery (or short circuit) current and the input voltage are not applied simultaneously to the transistor.

For most of the circuits (mainly non-isolated converters) where simultaneous high-voltage, high-current transients are applied to the transistors, the need for device protection will not be required, owing to the fact that the high speed switch is far more capable of absorbing this energy pulse in the forward biased mode than during turn-off. Nevertheless, the presence of such a power pulse will lead to higher average junction temperatures by way of dissipation in the switch.

When parasitic inductance is small and turn-on losses are considered excessive, a lumped series inductance is often purposely added in the form of a series turn-on snubber. The snubber reduces the rate of growth of the current $\dfrac{dI_c}{dt}$ for the turn-on time. It allows the reduction of turn-on losses in the transistor, and can lead indirectly to improved efficiency by allowing for a lower averaged junction temperature. Additionally, it may serve as a protective device in those situations where forward biased safe operating area may be jeopardised by an extremely severe turn-on load line.

Severe, and very often damaging, turn-on load lines can occur in half and full bridge PWM converter circuits fitted with the simple RCD turn-off switching aid network as described.

Protection Against Spike Voltages

The main purpose of the switching aid network at turn-off is to reduce the rate of rise of the re-applied voltage. Its effect as a spike voltage limiter is small.

If the presence of wiring inductance or leakage inductance of the transformer results in the generation of high spike voltages (fly back and push pull circuits in particular), it is possible to reduce these by using a component (eg. Zener diode) or a clamping circuit.

Parasitic Oscillations in MOSFETs

Power MOSFETs are very fast devices with appreciable gain at frequencies into the megahertz range. This high frequency gain, coupled with the internal and external circuit parasitic capacitances and inductances, particularly the loop inductance between discrete power MOSFETs, can cause unwanted parasitic oscillations to appear. The frequency of oscillation can range from 1 to 300 MHz.

The oscillations occur while the device is in the active mode where the transconductance is large. This means that in a switching application, the oscillations will occur on a transient basis during the turn-on and turn-off transitions. Parasitic oscillations cannot normally be tolerated and must therefore be eliminated. A small (1-5 ohm) non-inductive resistor in series with the gate lead of each device can reduce the problem to manageable proportions. A ferrite bead around individual gate leads is also sometimes used to good effect.

3.2 CONTROL REQUIREMENTS & TECHNIQUES

A power supply should be designed to:

(a) have good line regulation, such that the output remains constant if the input voltage varies;

(b) have good load regulation, such that the output remains constant if the load changes;

(c) have good transient response to system disturbances, such as sudden changes to the input voltage, or to the load;

(d) remain stable under all operating conditions.

The above requirements are met by designing a feedback control system, which will control the duty ratio, D, of the transistor to keep the output constant at all times.

A sudden increase in load (i.e. a reduction in R) will produce one of the following effects on the output voltage V_o .

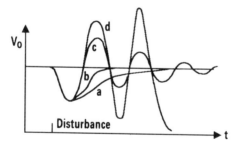

Output voltage response to an increase in load
Curve (a) over-damped (poor response)
Curve (b) critically damped (optimum)
Curve (c) under-damped (oscillatory)
Curve (d) unstable

A good power supply will return to the nominal output voltage very quickly, with minimal oscillations.

A simple buck regulator with feedback control is shown below, and underneath it is redrawn as a control block diagram.

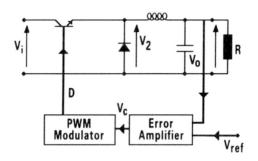

Buck regulator with feedback control

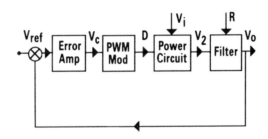

Power supply control block diagram

In designing the feedback loop, the transfer function of each block must be determined, and the loop transfer function optimised.

PWM CONTROLLERS

In most SMPS, the output voltage V_o is controlled by comparing the output voltage with a reference voltage, and using the resulting error to adjust the duty ratio D.

Normally, D is obtained by comparing the control voltage V_c with a fixed frequency sawtooth voltage, as illustrated below.

(a)

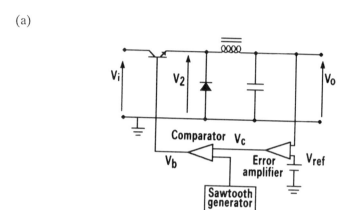

(b)

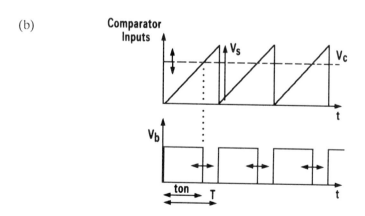

(a) Power supply with PWM control
(b) Control circuit waveforms

It can be seen that if the control voltage is increased, the duty ratio D is also increased. Thus:

$$D = \frac{t_{on}}{T} = \frac{V_c}{V_s}$$

Direct Duty Control

In this, the technique used in the majority of switched mode power supplies, the sawtooth waveform has a fixed amplitude as well as fixed frequency, and the duty ratio D is adjusted only by a change in the output voltage altering the control voltage V_c .

This technique is popular as it is very simple, but it does have drawbacks. The open loop line regulation is poor, ie. if the feedback loop is broken, a change in V_i will cause a significant change in the output voltage V_o .

Direct duty control also has a poor closed loop transient response to changes in V_i , as a change in V_i will result in a compensating change in D delayed by the output L-C filter.

Voltage Feedforward Control

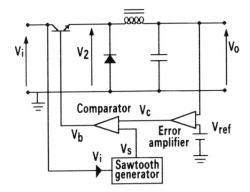

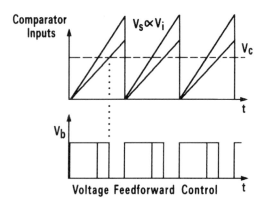

Buck regulator with voltage feedforward control

This is very similar to direct duty control, except that the magnitude of the sawtooth waveform, V_s , is proportional to the input voltage V_i .

Therefore:

$$V_s = \frac{V_i}{K}$$

where K is a constant.

But:

$$D = \frac{V_c}{V_s}$$

Therefore:

$$D = K.\frac{V_c}{V_i}$$

In a buck regulator operating in the continuous mode, the d.c. gain is:

$$\frac{V_o}{V_i} = D$$

$$= \frac{K.V_c}{V_i}$$

Therefore:

$$V_o = K.V_c$$

Thus the output is no longer dependent on the input voltage, ie. the system has excellent open loop line regulation.

The closed loop transient response to input voltage changes is also excellent, as the delay caused by the L-C filter is not in the feedforward control line. Thus the compensating change in D is instantaneous and precise.

The d.c. gain for the 3 basic regulators in both modes of operation is shown below:

(a) Buck, Continuous.

$$\frac{V_o}{V_i} = D$$

(b) Buck, Discontinuous.

$$\frac{V_o}{V_i} = \frac{2D}{D + \sqrt{D^2 + (8L/RT)}}$$

(c) Flyback, Continuous.

$$\frac{V_o}{V_i} = \frac{D}{1 - D}$$

(d) Flyback, Discontinuous.

$$\frac{V_o}{V_i} = D. \sqrt{\frac{RT}{2L}}$$

(e) Boost, Continuous.

$$\frac{V_o}{V_i} = \frac{1}{1-D}$$

(f) Boost, Discontinuous.

$$\frac{V_o}{V_i} = \frac{1 + \sqrt{1 + 2D^2 TR/L}}{2}$$

By substituting $D = (K.V_c)/V_i$ into the above equations, it can be seen that V_i disappears from the d.c. gain in only the buck (continuous mode) and the flyback (discontinuous mode) regulators. In all other regulators voltage feedforward provides only partial compensation, and is rarely used. However, as the large majority of SMPS are continuous mode buck or discontinuous mode flyback, voltage feedforward control is a very important and easily implemented technique.

It should be noted that voltage feedforward is an open loop technique; therefore it is important that the scaling factors are designed correctly.

Current Mode Control

This technique was first reported in 1977, and its use has become widespread since.

Instead of comparing the control voltage to an independently generated sawtooth voltage, the V_c is compared to a voltage derived from the inductor current, forming a second, inner control loop.

In the buck regulator shown below, the clock sends a pulse to the R-S flip-flop which sets the Q output to logic "1", thus switching on the main transistor. The inductor current will then build up in a linear fashion:

$$V_i - V_o = L.\frac{dI_L}{dt}$$

When V_s ($= I_L.r$) reaches the control voltage V_c , the comparator output changes to a logic "1", resetting the flip-flop Q output to "0", hence switching off T_1. V_s immediately goes to zero, and the comparator output reverts to logic "0".

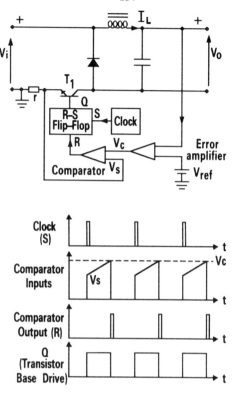

Buck regulator with current mode control

As with voltage feedforward control, current mode control eliminates the dependence of V_o on input voltage V_i. If the input voltage V_i increases, then the inductor voltage $(V_i - V_o)$ increases, and the gradient dI_L/dt increases. Thus V_s reaches V_c more quickly, reducing t_{on} and hence the duty ratio D. There are no delays in the loop, so duty ratio compensation is instantaneous and precise. Thus open loop line regulation with current mode control is excellent, as is the closed loop transient response to input voltage changes.

With current mode control, if V_o reduces, then the control voltage V_c will increase, delaying the comparator reset pulse and hence increasing the duty ratio D. However, a reduction in V_o will increase the gradient of I_L, and the result of these two effects gives an easily stabilised system.

Current limit protection is easily implemented, by limiting the maximum value of V_c and hence I_L (a zener diode across V_c would give an effective, though not adjustable, clamp). In addition, precise sharing of the load current between two (or more) SMPS can be achieved by feeding a common V_c into the comparator of each SMPS.

The main disadvantage of current mode control is that the control is dependent on the load current - a higher value of V_c and hence steady state error is apparent at high load currents.

ISOLATION IN THE FEEDBACK LOOP

The circuits discussed in Section 1.3 provide electrical isolation between the input and the output, so that the output is floating. However, as soon as the feedback loop is closed, there is another connection between the input and the output, as shown below.

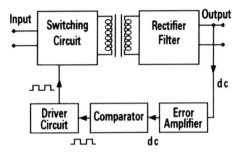

Block diagram of a power supply with isolation

For the output to be floating, some form of isolation is necessary in the feedback loop. This can be provided by magnetic coupling, using a transformer, or, since the feedback loop is low power, opto-coupling can be used.

Magnetic Coupling

Transformers only work with a.c., so a transformer cannot be inserted directly between the output and the error amplifier, or between the error amplifier and the comparator. If a transformer is to be inserted here, a high frequency a.c. carrier waveform would have to be generated, and the d.c. signal used to amplitude modulate the carrier. This is sometimes done, but it adds considerably to the complexity of the circuit.

Much more common is to insert a pulse transformer between the driver circuit and the switching transistor(s). This may be included anyway, particularly in ½-bridge and full bridge circuits, where the transistor bases/gates are referenced to different voltages levels. However, this produces another problem, as the control and driver circuits are connected to the secondary, and therefore must derive their power from the secondary. This can create start-up problems, as there will be no secondary voltage until the transistors have started switching. This will be discussed in more detail in Section 3.3.

Opto-Coupling

Another alternative is to use an opto-isolator, either between the output and the error amplifier, or between the error amplifier and the comparator. This means that the control and driver circuits can be powered from the d.c. input (which may cause problems if operating from a high d.c. voltage). However, opto-isolators tend to be very non-linear, and extra biasing circuitry may be necessary to produce a reasonably linear transfer function. This is not a major problem if the opto-isolator is placed after the error amplifier, as any non-linearities are compensated for by the feedback loop.

In circuits where regulation is not particularly tightly specified, the feedback signal can be taken from an extra winding on the transformer. This is the simplest technique, but no feedback compensation will be made for any volt drops on the secondary circuit, such as across the rectifying diodes or the filter inductor.

POWER SUPPLIES WITH MULTIPLE OUTPUTS

In many applications, there is more than one output, each fed from a separate transformer secondary. The output voltage feedback signal, which is fed into the control circuit to set the transistor duty ratio D, can only be taken from one output. Hence the value of D will be set at a level which will keep that particular output at the required voltage, regardless of the voltage at the other outputs.

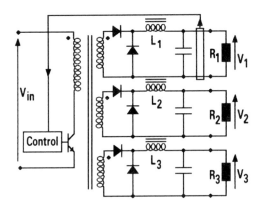

Forward converter with 3 outputs

In the case of the above circuit operating in the continuous mode, the d.c. gain is:

$$\frac{V_o}{V_i} = 2.n.D$$

$$= 2.n.D. \frac{R}{R + r_L}$$

including the effect of inductor resistance, r_L.

From the above, it can be seen that a change in the input voltage V_i would affect each output in the same proportion, so the feedback signal from output 1 would produce an adjustment in the duty ratio D which would correct each output. Hence the *line* regulation for the auxiliary outputs would be approximately the same as for the main output.

In a practical converter where $r_L \neq 0$, a change in load resistance R will produce a change in the output voltage. In the case of the main output, the feedback signal will produce a compensating adjustment in D. However, a change in load in any of the auxiliary outputs will not substantially affect the main output voltage, therefore there will be no compensating adjustment in D. Thus the *load* regulation of the auxiliary outputs will not be as good as for the main output. This will be accentuated if discontinuous mode operation is used, as even in the ideal situation of $r_L = 0$, the output voltage depends on the load R.

Typically, a multiple output SMPS may have 0.5% load regulation on the main output, but only 3% on each of the auxiliary outputs.

If close regulation on an auxiliary output is required, then a post-regulator must be used. This is normally either a linear regulator or a basic buck regulator, which will add considerably to the size and weight of the total power supply. Magnetic amplifiers are also making a comeback as post-regulators, due to the development of new core materials and custom magnetic amplifier control integrated circuits.

Dynamic cross-regulation between the regulated output and the unregulated outputs tends to be poor, as it takes a long time for the control circuit to see the effect on the regulated output of a load change on an unregulated output. A useful technique for improving the dynamic cross-regulation is to wind the output inductors on the same core, so that a change in an unregulated output is immediately coupled to the regulated output, which can be then corrected by the feedback circuit. However, this can make it more difficult to stabilise the power supply.

3.3 CONTROL CIRCUITRY

This section will deal with a typical SMPS control integrated circuit and then consider the problem of powering the control circuitry. The designer of a SMPS has a wide range of integrated circuit controllers available from different semiconductor manufacturers. The decision whether to use an integrated circuit, or discrete circuitry, will depend on factors such as cost, ease of assembly and noise immunity. Integrated circuits are convenient for low and medium power SMPS, but they tend to have relatively poor amplifiers and poor noise immunity, which makes them less useful in high powered SMPS applications.

The UC3842 is a popular low-cost IC current-mode controller for SMPS and is typical of a wide range of such devices that are available. It comes in an eight pin DIL package. The figure below shows its block diagram.

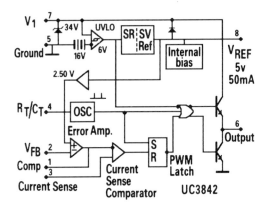

Block diagram of UC3842

The circuit blocks in the top half of the diagram are to provide the necessary internal regulated power rails for the circuitry. They also provide start-up sequencing to allow the IC to have a low start-up current of around 1mA. This makes it easier to power in a practical SMPS circuit for reasons that will be discussed in the next section. There is also an undervoltage lock out (UVLO) facility which shuts down the whole power supply when the input voltage (V1) drops below 16 volts.

Pins 5 and 7 are the power supply for the IC itself, which should lie in the range 16 to 34 volts. Pin 8 provides a 5 volt reference, capable of providing up to 50mA. Pin 4 has the timing resistor, R_t, and capacitor, C_t, attached to it. These components are used to set the switching frequency of the power supply. R_t is connected from pin 8 to pin 4 and C_t from pin 4 to ground. C_t charges up through R_t, from the 5 volt rail, and is then discharged by the oscillator circuitry in the IC. This cycle then repeats itself *ad infinitum*.

The sampled output voltage of the power supply is connected to pin 2 and the sampled output current is fed to pin 3. Pin 1 is to provide the output of the error amplifier and is for adding compensation feedback to optimise the response of the power supply to changes in load conditions. The S-R latch is used to vary the duty cycle of the switching waveform in response to differing loads. The output from the PWM latch is used to drive the totem pole output on pin 6. This output configuration is ideal for driving the capacitance of a MOSFET gate.

POWERING THE CONTROL CIRCUITRY

The provision of power for the control electronics in a SMPS can present a problem. It is not possible to take power from the regulated output, as the control circuitry must come up before the output voltage can build up. This means that power for the control circuit must be derived directly from the mains input and not via the control loop.

One solution to this problem is to provide a dropping resistor from the rectified mains supply to give a low voltage supply for the control electronics. This has the disadvantages that the circuitry is not isolated from the mains and that considerable power can be dissipated in the dropper resistor. In some applications, where low cost and simplicity are the over-riding considerations this solution may be the most appropriate.

Another possibility is to use a low voltage step-down transformer from the 50Hz mains. This is a relatively expensive solution, which adds weight and bulk to a design. Remember that the 50Hz transformer may be nearly as large as the main switching transformer because of the frequency difference, despite the fact that it is only powering the control circuitry. This solution does have the advantage that it provides mains isolation for the control electronics.

In aerospace applications where the mains frequency is normally 400Hz this solution is more attractive, as the size of the transformer is considerably reduced.

A third method that can be used to power the control electronics is shown below.

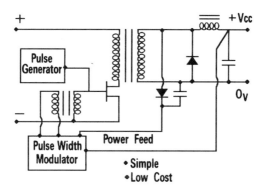

An alternative method of powering the control electronics

At initial switch-on the pulse-width modulator will not be running as its power supply will not be present. The pulse generator is used to supply a series of pulses to switch the MOSFET, allowing the main transformer to transmit sufficient power to start powering up the PWM, which then takes over the control of the power supply. The pulse generator can be powered directly from the mains as it does not require isolation. Usually the pulses are derived directly from the mains a.c.

waveform. This scheme gives isolation and can be implemented at little cost. It is worth pointing out that it is best to use an IC like the UC3842, which has a low start-up current, for this scheme to work well.

3.4 BODE DIAGRAMS

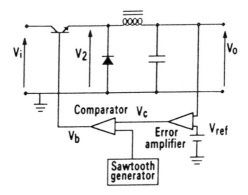

A typical SMPS including the feedback control circuit

The above feedback circuit must be designed such that the system:

(a) has good line regulation,

(b) has good load regulation,

(c) has good transient response to system disturbances,

(d) remains stable under all operating conditions.

Possible system disturbances are:

(a) change in input voltage, V_i ,

(b) change in load resistance, R,

(c) Change in V_{ref} - never likely to occur.

As discussed in Section 3.2, the effects of a change in V_i can be largely eliminated by the use of either voltage feedforward control or current mode control in continuous mode buck and discontinuous mode flyback regulators.

A sudden increase in load (reduction in R) will produce one of the following effects on the output voltage V_o :

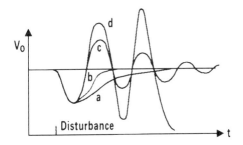

Curve   (a)   over-damped (poor transient response)
        (b)   critically damped (optimum)
        (c)   under-damped (oscillatory)
        (d)   unstable

Effect of sudden increase in load current on output voltage

The block diagram of a SMPS system is shown below.

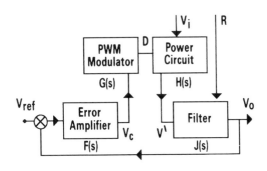

SMPS block diagram

The transfer function of the PWM (Pulse Width Modulator), the power circuit, and the filter are fixed by other considerations discussed earlier (eg. L is determined by the continuous current mode limit and by the peak inductor current, C by the maximum ripple voltage), and cannot be altered. However, a potentially unstable loop can be stabilised by adding suitable compensation components to the error amplifier.

BODE PLOTS

Bode's technique of plotting the system gain and phase response, for a sinusoidal disturbance, against frequency, provides a very useful graphical method of analysing the system and calculating the necessary compensation components to ensure both stability and a fast transient response.

R-C Low-pass Network

$$V_i \quad R \quad C \quad V_0 \quad X_C = \frac{1}{j\omega C}$$

R-C low-pass network

In the above circuit, for a sinusoidal input V_i at frequency ω ($=2\pi f$):

$$V_o = V_i \times \left[\frac{\dfrac{1}{j\omega C}}{R + \dfrac{1}{j\omega C}} \right]$$

Therefore:

$$\frac{V_o}{V_i} = \frac{1}{1 + j.RC.\omega}$$

$$= \frac{1}{1 + j.\dfrac{\omega}{\omega_p}}$$

where:

$$\omega_p = \frac{1}{RC}$$

If $\omega \ll \omega_p$

$$Gain = \frac{V_o}{V_i} = 1$$

or, expressed in decibels:

$$Gain = 20 \log_{10}\left(\frac{V_o}{V_i} \right)$$

$$= 20 \log_{10}(1)$$

$$= 0 \; dB$$

The phase difference between V_i and V_o is $0°$.

If $\omega \gg \omega_p$:

$$\frac{V_o}{V_i} = \frac{\omega_p}{\omega}$$

Therefore, the gain in decibels is:

$$Gain = 20 \log_{10} \left(\frac{\omega_p}{\omega} \right)$$

$$= 20 \log_{10} \omega_p - 20 \log_{10} \omega$$

$$(= 0 \ dB \qquad at \ \omega = \omega_p \ )$$

The slope of gain (dB) against frequency is -20 dB/decade.

Also, V_o lags V_i by 90°.

At $\omega = \omega_p$, V_o lags V_i by 45°.

The circuit gain and phase can be plotted against frequency on log-linear graph paper.

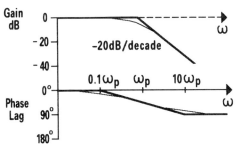

Gain and phase response of R-C low-pass network

If $s = j\omega$ then:

$$\frac{V_o}{V_i} = F\left(s\right) = \frac{1}{1 + j\dfrac{s}{\omega_p}}$$

The denominator $= 0$ when $s = -\omega_p$

ie. a *pole* exists at ω_p $\left(F\left(s\right) \to \infty\right)$

A network with opposite gain and phase responses is said to introduce a zero into a system. In the Bode diagram on the following page we see the gain and phase responses of a network with a single zero. The gain response increases at 20dB per decade, from ω_o, while the phase response has a 90° lead centred on ω_o. Such a network can be useful to compensate the response of a system with several poles.

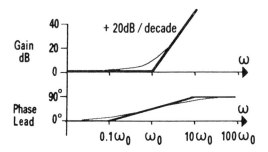

The gain and phase response associated with a zero

To summarise: a pole has a phase lag and a down-turned frequency response plot, while a zero has a phase lead and an up-turned frequency response plot associated with it.

The L-C Network

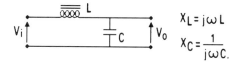

The L-C network

Ideally, for a lossless circuit:

$$\frac{V_o}{V_i} = \frac{\left(\dfrac{1}{j\omega C}\right)}{j\omega L + \left(\dfrac{1}{j\omega C}\right)}$$

$$= \frac{1}{1 + (j\omega)^2 LC}$$

$$= \frac{1}{1 + \dfrac{(j\omega)^2}{(\omega_p)^2}}$$

where:

$$\omega_p = \frac{1}{\sqrt{LC}}$$

In the above ideal, lossless, circuit, V_o/V_i tends to infinity at $\omega = \omega_p$.

The effective circuit resistance R_s limits the gain at $\omega = \omega_p$ to a finite value, depending on the circuit Q factor, where:

$$Q = \frac{\omega L}{R_s}$$

The gain becomes:

$$\frac{V_o}{V_i} = F(s) = \frac{1}{1 + \left[\dfrac{s}{\omega_p} \cdot \dfrac{1}{Q}\right] + \left[\dfrac{s^2}{\omega_p^2}\right]}$$

which has a double pole at $\omega = \omega_p$.

The gain and phase plots against frequency are shown below for different Q values.

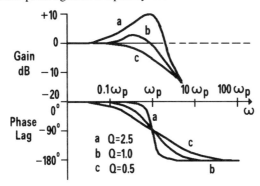

Gain and phase plots for different Q values

Blocks In Cascade

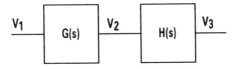

Two cascaded networks

$$V_2(s) = G(s).V_1(s)$$
$$V_3(s) = H(s).V_2(s)$$

Therefore:

$$\frac{V_3(s)}{V_1(s)} = G(s).H(s)$$

ie. The gains of cascaded blocks multiply.

Plotting Functions With Several Poles And Zeros

Example:

$$G(s) = \frac{10 \cdot (1 + 0.1s)}{s \cdot (1 + 0.5s)}$$

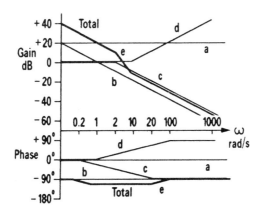

A function with a constant gain term, two poles and a zero

(a) Constant gain term, (10):

In decibels, $= 20 \log_{10} 10 = 20 \ dB$

No effect on phase.

(b) First pole term, $\left(\dfrac{1}{s}\right)$:

In decibels, $= 20 \log_{10} \left(\dfrac{1}{s}\right)$

This is a straight line of gradient -20 dB/decade, passing through 0 dB at $\omega = 1$ rad/s.

It will produce a 90° phase lag at all frequencies.

(c) Second pole term, $\left(\dfrac{1}{1 + 0.5s}\right)$:

In decibels, $= 20 \log_{10} \left(\dfrac{1}{1 + \dfrac{s}{2}}\right) = -20 \log_{10} \left(1 + \dfrac{s}{2}\right) \ dB$

This will be 0 dB from 0 to s=2 rad/s, thereafter a straight line of gradient -20 dB/decade.

There will be 0° phase lag from 0 to 0.2 rad/s,
a 90° lag from 20 rad/s upwards, and
a straight line from 0.2 rad/s to 20 rad/s, passing through 45° at 2 rad/s.

(d) Zero term, $(1 + 0.1s)$:

$$\text{In decibels,} = 20 \log_{10} \left(1 + \frac{s}{10}\right) \text{ dB}$$

This will be 0 dB from 0 to 10 rad/s, thereafter a straight line of gradient +20 dB/decade.

There will be 0° phase lag up to 1 rad/s,
a 90° phase *lead* from 100 rad/s,
a straight line from 1 rad/s to 100 rad/s, passing through 45° lead at 10 rad/s.

STABILITY ANALYSIS

(a) Stability

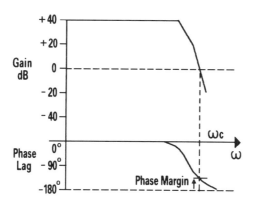

A typical Bode diagram showing phase margin

For a stable system, the phase lag at ω_c (the frequency where the gain passes through 0 dB) must be less than 180°.

If the *Phase Margin* (=180° - phase lag at ω_c) is small, the system will be very oscillatory. A phase margin of 45° generally gives a good response.

(b) Transient Response

A fast transient response will be achieved if the value of ω_c is kept high.

(c) Regulation

Good *closed* loop line and load regulation will be achieved if the d.c. gain (the gain at $\omega =0$) is high.

3.5 ERROR AMPLIFIER COMPENSATION TECHNIQUES

The block diagram of a switched mode power supply as a control system is shown below.

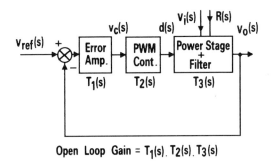

Open Loop Gain = $T_1(s)$. $T_2(s)$. $T_3(s)$

Power supply control block diagram

The transfer function of the power stage and filter, $T_1(s)$, depends on the circuit topology chosen, and whether it is operating in the continuous or discontinuous current mode. $T_2(s)$ depends on the type of controller used: normally direct duty, voltage feedforward, or current mode. There is little flexibility to alter either T_2 or T_3; however, the performance of the complete power supply can be optimised by careful design of the error amplifier transfer function, T_1.

POWER CIRCUIT AND FILTER

Power supplies operating in the continuous current mode have a dominant 2-pole characteristic associated with the capacitor and the inductor, plus a (high frequency) zero from the capacitor ESR.

The inductor current when load is suddenly applied to a buck converter operating in the continuous current mode is shown below:

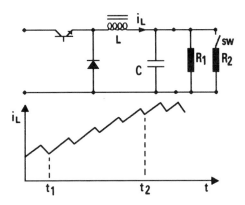

Inductor current in a continuous mode buck regulator as load is rejected

It takes several cycles for the inductor current to build up to the new load current, due to the large value of inductance necessary to maintain continuous current.

If, however, the circuit is designed to operate in the discontinuous current mode, then the inductance has to be small enough for the current to be able to build up to any value within a cycle, as shown below.

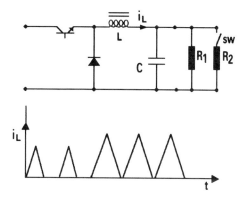

Inductor current in a discontinuous mode buck regulator as load is rejected

As changes in the duty ratio D can only be made once per cycle, the inductor disappears from the feedback loop, leaving a single pole characteristic associated with the filter capacitor, which is much easier to stabilise.

The d.c. gain of the flyback regulator operating in the continuous current mode is:

$$\frac{V_o}{V_i} = \frac{D}{(1-D)}$$

and for the boost regulator operating in the continuous current mode:

$$\frac{V_o}{V_i} = \frac{1}{(1-D)}$$

The term $(1-D)$ in the denominator has a dramatic effect on the response of the circuits to sudden load changes.

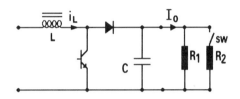

Boost converter

In the boost converter above, the output voltage will drop, if load R_2 is switched on. The feedback circuit will compensate by increasing D to bring V_o back to its rated value. The inductor has a large value in continuous mode circuits, so i_L cannot quickly increase to the new load current, so the effect of increasing D is only to reduce the time current is flowing through the diode, so *reducing* the energy transferred, causing the output voltage to *reduce* still further.

The term $1/(1-D)$ leads to a *right half plane zero* in the open loop transfer function. The effect of an RHP zero is, like a normal LHP zero, to reduce the gain slope by -20 dB/decade, but it *increases* the phase lag by 90°. This makes continuous mode flyback and boost circuits very difficult to stabilise, and for this reason flyback and boost regulators are usually operated in the discontinuous current mode.

The two most common circuit topologies used are continuous mode buck converters and discontinuous mode flyback converters.

Continuous Mode Buck Converters

The d.c. gain of a continuous mode buck converter is:

$$\frac{V_o}{D} = V_i$$

Under small signal a.c. conditions with either direct duty or voltage feedforward the transfer function is:

$$T_3(s) = \frac{v_o(s)}{d(s)} = V_i.H_e(s)$$

where:

$$H_e(s) = \frac{\left(1 + \dfrac{s}{\omega_z}\right)}{1 + \left(\dfrac{s}{\omega_p}.\dfrac{1}{Q}\right) + \left(\dfrac{s^2}{\omega_p^2}\right)}$$

with:

$$\omega_z = \left(\frac{1}{R_c.C}\right)$$

$$\omega_p = \frac{1}{\sqrt{L.C}}$$

$$Q = \left(\frac{R}{\omega_p.L}\right)$$

R_c = capacitor effective series resistance (ESR)
R = load

The asymptotic plot of the gain and phase of $T_3(s)$ against frequency are shown below. It should be noted that the Q-factor of the combined filter and load can cause peaking at the double pole, which may result in the response differing markedly from the asymptotic plot around this double pole.

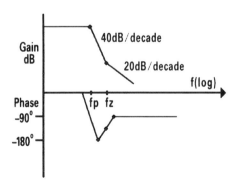

Bode plot of a continuous mode buck converter

There is a double pole at f_p, the filter resonant frequency, and a zero associated with the capacitor ESR.

Discontinuous Mode Flyback Converters

The d.c. gain of a flyback converter is:

$$\frac{V_o}{D} = V_i \cdot \sqrt{\frac{R}{2.L.f}}$$

Under small signal a.c. conditions:

$$T_3(s) = \frac{v_o(s)}{d(s)} = V_i \cdot \sqrt{\frac{R}{2.L.f}} \cdot H_e(s)$$

where:

$$H_e(s) = \frac{\left(1 + \dfrac{s}{\omega_z}\right)}{\left(1 + \dfrac{s}{\omega_p}\right)}$$

with:

$$\omega_z = \left(\frac{1}{R_c.C}\right)$$

$$\omega_p = \left(\frac{2}{R.C}\right)$$

The gain and phase of $T_3(s)$ are plotted against frequency below:

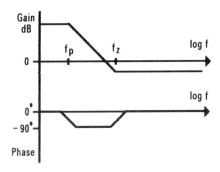

Bode plot of a discontinuous mode flyback converter

There is a single pole associated with the capacitor and the load, and a high frequency zero associated with the capacitor ESR.

PWM CONTROLLER

(a) Direct Duty

$$T_2(s) = \frac{d(s)}{v_c(s)} = \frac{1}{V_s}$$

where V_s is the peak sawtooth voltage.

(b) Voltage Feedforward

$$T_2(s) = \frac{d(s)}{v_c(s)} = \frac{K}{V_i}$$

(c) Current Mode

Current mode control removes the inductor pole from the power stage and filter transfer function, so it is easier to consider the combined transfer function $T_2(s).T_3(s)$. In current mode control:

$$I_L = K.V_c$$

In buck regulators, $I_L = I_o$, so under d.c. conditions:

$$V_o = I_o.R$$
$$= K.V_c.R$$

Under small signal a.c. conditions:

$$\frac{v_o(s)}{v_c(s)} = K.R.H_e(s)$$

where:

$$H_e(s) = \frac{1 + \dfrac{s}{\omega_z}}{1 + \dfrac{s}{\omega_p}}$$

with:

$$\omega_p = \frac{1}{R.C}$$

$$\omega_z = \frac{1}{R_c.C}$$

In discontinuous mode flyback circuits:

$$I_o = I_L . D_1$$

where:

$D_1 =$ *duty ratio of the series diode*

$$= \sqrt{\frac{2L}{RT}}$$

So:

$$V_o = I_o.R$$

$$= K.V_c \cdot \sqrt{\frac{2LR}{T}}$$

using $I_L = K.V_c$.

The small signal transfer function is:

$$\frac{v_o(s)}{v_c(s)} = K.V_c \cdot \sqrt{\frac{2LR}{T}} \cdot H_e(s)$$

where:

$$H_e(s) = \frac{1 + \dfrac{s}{\omega_z}}{1 + \dfrac{s}{\omega_p}}$$

with:

$$\omega_p = \frac{2}{R.C}$$

$$\omega_z = \frac{1}{R_c.C}$$

ERROR AMPLIFIER

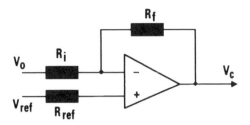

Error amplifier with proportional feedback control

In the error amplifier shown in the system above:

$$V_c = -\frac{R_f}{R_i}.(V_o - V_{ref}) + V_{ref}$$

Therefore the gain of the amplifier is (R_f/R_i), or in decibels $20 \log_{10}(R_f/R_i)$. This is a constant value, independent of frequency, producing zero phase shift.

By adding suitable compensating resistors and capacitors around the error amplifier, its gain and phase frequency response can be shaped to ensure stability under all conditions, whilst retaining good output voltage regulation and a fast transient response.

There are a number of standard compensation circuits, each with well documented characteristics. Two of these, which between them are suitable for the majority of SMPS circuits, are given below.

(a) Single Pole Compensation

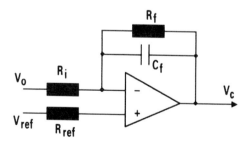

Error amplifier with single pole compensation

The low frequency gain $\left(R_f \ll \dfrac{1}{\omega C_f} \right)$ is the same as in the uncompensated circuit shown previously, (R_f/R_i) ; or, in decibels:

$$Gain = 20 \log_{10} \left(\frac{R_f}{R_i} \right) \quad dB$$

The phase lag here is $0°$.

For $R_f >> \dfrac{1}{\omega C_f}$, the gain becomes:

$$Gain = 20 \log_{10} \left[\frac{\left(\dfrac{1}{\omega C} \right)}{R_i} \right] = -20 \log_{10} (K. \omega)$$

ie. a slope of -20 dB/decade.

The phase lag is now $90°$.

The single pole (break point) is at $\omega = \omega_p = \dfrac{1}{R_f . C_f}$

R_{ref} should be equal to R_i to cancel out the amplifier input voltage offset caused by the amplifier input bias current.

(b) Two Pole Compensation

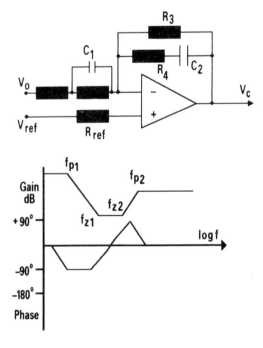

Error amplifier with two pole compensation

$$d.c. \ gain = \left(\frac{R_3}{R_1 + R_2} \right)$$

The poles occur at:

$$\omega_{p1} = \frac{1}{(R_3 + R_4).C_2}$$

which tends to zero if R_3 is omitted (as it sometimes is).

$$\omega_{p2} = \left(\frac{R_1 + R_2}{R_1.R_2.C_1} \right)$$

The zeros occur at:

$$\omega_{z1} = \frac{1}{R_4.C_2}$$

$$\omega_{z2} = \frac{1}{R_2.C_1}$$

Usually the two zeros are designed to be at the same frequency, ie:

$$\omega_{zi} = \omega_{z2}$$

This circuit does not have a good response to large disturbances, due to saturation of capacitor C_1.

Single Pole Compensation circuits are used for SMPS where the filter has a single pole characteristic. These are:

1. all discontinuous mode circuits;
2. all continuous mode circuits using current mode control.

In all these circuits the inductor current is controlled on a cycle-by-cycle basis (it does not depend on the current in the previous cycle), and therefore the inductor does not constitute a delay in the feedback circuit.

Two Pole Compensation circuits are used for SMPS where the filter has a two pole characteristic. This is all continuous mode circuits using direct duty or voltage feedforward control.

It should be noted that continuous mode flyback and continuous mode boost circuits usually require extra stabilising components due to the term $\left(\frac{1}{1 - D} \right)$ in the gain.

Crossover Frequency

Switched mode power supplies are essentially sampled data systems, therefore the maximum theoretical crossover frequency is $f_c = \left(\dfrac{f_s}{2} \right)$ (Nyquist sampling theorem). It is usual to design the feedback loop with a crossover frequency between $\left(\dfrac{f_s}{10} \right)$ and $\left(\dfrac{f_s}{5} \right)$, which should ensure stability, while still providing a fast transient response.

$f_s \quad = \quad$ switching frequency

$f_c \quad = \quad \left(\dfrac{\omega_c}{2\pi} \right)$

EXAMPLE CALCULATION

SMPS Specification

Flyback regulator, discontinuous mode, direct duty control.

Frequency = 100 kHz (ie. T = 10 microseconds).

Input: 10 to 15 volts.

Output: -15 volts, 1 amp to 5 amps.

L = 3 μH

C = 10,000 μF

R_c = 15 mΩ

V_s = 2.5 volts

(a) D.C. Conditions

Direct duty PWM modulator:

$$\frac{D}{V_c} = \frac{1}{V_s}$$

Power circuit, flyback, discontinuous mode:

$$\frac{V_o}{V_i} = D. \sqrt{\frac{RT}{2L}}$$

Therefore:

$$\frac{V_o}{V_c} = V_i \cdot \sqrt{\frac{R.T}{2.L}} \cdot \frac{1}{V_s}$$

So:

$$\frac{V_o}{V_c} = 26 \ dB \quad \text{for} \ V_i = 10, \ R = 15 \ \Omega$$

$$= 19 \ dB \quad \text{for} \ V_i = 10, \ R = 3 \ \Omega$$

$$= 30 \ dB \quad \text{for} \ V_i = 15, \ R = 15 \ \Omega$$

$$= 23 \ dB \quad \text{for} \quad V_i = 15, \ R = 3 \ \Omega$$

(b) Small Signal Transfer Function

$$\frac{v_o(s)}{v_c(s)} = \frac{V_i}{V_s} \cdot \sqrt{\frac{R.T}{2.L}} \cdot \frac{\left(1 + \dfrac{s}{\omega_z}\right)}{\left(1 + \dfrac{s}{\omega_p}\right)}$$

where

$$f_p = \frac{2}{2.\pi.R.C}$$

$$= 10 \ Hz \quad \text{for} \quad R = 3 \ \Omega$$

$$= 2 \ Hz \quad \text{for} \quad R = 15 \ \Omega$$

$$f_z = \frac{1}{2.\pi.R_c.C}$$

$$= 1060 \ Hz$$

(c) Error Amplifier Compensation

As the discontinuous mode flyback regulator has a single pole characteristic, the single pole error amplifier will be used.

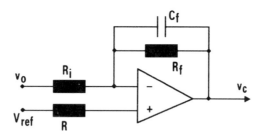

Error amplifier with single pole compensation

Crossover frequency, $f_c = \dfrac{f_s}{10} = 10 \ \text{kHz}$

A pole to compensate for the filter capacitor zero is placed at

$$f_{p2} = \frac{f_z}{10} = 100 \ Hz$$

thus increasing the overall phase lag to $(90° + 45°)$, but still leaving a phase margin of $45°$.

The overall plot can now be drawn, passing through the point (10 kHz, 0 dB), with the gradient reducing by 20 dB/decade at every pole, and increasing by 20 dB/decade at every zero.

The overall curve is drawn for case (a), as this is the most critical.

The required error amplifier d.c. gain = 80 - 23 = 57 dB.

Therefore:

$$20 \log_{10} \left(\frac{R_f}{R_i} \right) = 57$$

ie:

$$R_f = 741 \times R_i$$

If $\quad R_i = 3\,k\Omega$
$\quad\quad R_f = 2.2\,M\Omega$

The error amplifier pole is at 100 Hz, therefore:

$$f_{p2} = 100 = \frac{1}{R_f.C_f}$$

Therefore:

$$C_f = 723\,pF$$

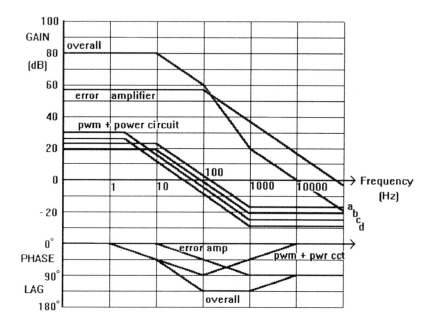

Complete Bode plot

SECTION 4

PRACTICAL SMPS
DESIGN CONSIDERATIONS

4.1 ELECTROMAGNETIC INTERFERENCE (EMI) SUPPRESSION

Switched mode power conversion has many advantages making it very desirable, or even essential in some applications. Its one major drawback is the generation of high frequency electrical noise associated with the fast switching waveforms in the power regulator. In most applications it is necessary to filter out this noise and prevent it radiating from the power supply by the use of screens.

EMI is generated by rapidly changing electric or magnetic fields. Common sources of EMI are electric motors (particular commutator motors), relays and switches where rapid changes of current flow produce a broad range of interference frequencies.

As switched mode power supplies are continually switching current, they are potentially major sources of EMI (linear supplies generate much less). The square wave output of a switched mode supply is designed to have very steep rise and fall times to minimise the transistor switching losses, but the steeper the edges the greater the harmonic content, and the greater the likelihood of EMI being a problem.

It is necessary that switched mode power supplies are designed to minimise EMI and that they can operate in close proximity with other electronic equipment, without interfering with its operation. They should also be resistant to interference from external or internal sources, in other words they should be Electromagnetically Compatible (EMC) with other systems.

To see how EMI can arise, consider the following simple switching circuit. Initially with the switch open there is an electric field only around the wire connected to the battery. This is a static field and does not induce interference in surrounding circuits.

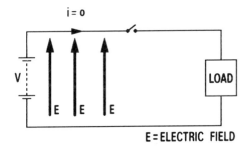

Simple switching circuit with switch open

When the switch is closed magnetic fields are established around the conductors and an electric field is established in the right hand half of the circuit.

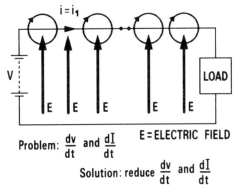

Simple switching circuit with switch closed

The establishment of the field flux in the space around the wires involves rapid changes of these fields. These changing or moving fields may intersect other conductors in nearby circuits, inducing unwanted voltages and currents, namely interference. The amount of interference produced by the circuit will depend on the rate of change of voltage and current in the circuit and also upon the size or geometry of the circuit. A large, spread out, circuit will produce more interference than a small compact one for a given current or voltage change.

In the simple switched mode power supply circuit shown below it is possible to identify the areas that give rise to EMI by considering the changes in voltage and current in the circuit.

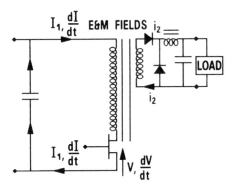

Simple SMPS circuit showing sources of EMI

The area around the switching transistor has high values of $\dfrac{dV}{dt}$ associated with it, while the conductors from the reservoir capacitor have high $\dfrac{dI}{dt}$ values. Both can provide severe interference if appropriate steps are not taken.

Another possible source of interference is the circuitry associated with the rectifier diodes. Here the rectification action of the diodes gives rise to discontinuous changes in current and hence high $\frac{dI}{dt}$ values.

EMI can be transmitted by electrical conductors, electric coupling (capacitive) or magnetic coupling (inductive) or by free space radiation. The figure below shows the potential EMI problem areas in a switching supply.

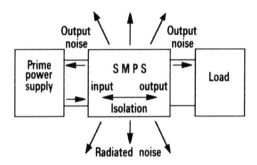

The potential EMI problem areas in a SMPS

They are:

- Noise output on the output connections to the load;

- Noise output on the input power connections;

- Direct free space radiation.

- Additionally the supply must provide isolation between its input and output and not pass noise from the prime power supply onto the output and vice-versa.

For EMI to be a problem in any electronic system there are three requirements.

- There must be a source of interference.

- There must be a transmission medium.

- There must be a susceptible receiver.

Removing any one of these will cure the problem, but ideally all three should be eliminated to avoid unexpected problems when the equipment is in use.

In order to solve an EMI problem it is important to realise that there two ways in which noise is transmitted or propagates in an equipment. It can be:

- Conducted noise carried along wiring;

- Radiated noise transmitted through space between wires.

To eliminate a noise problem there are three steps that should be taken. Firstly the noise should be reduced at source by reducing the values of $\frac{dV}{dt}$ and $\frac{dI}{dt}$ in the circuit if possible. Attention should also be paid to reducing stray electric and magnetic fields by careful layout. Secondly, conducted noise paths should be broken using frequency filters and finally radiation should be contained by the use of screens. In the next section consideration will be given to the implementation of these measures.

EMI REDUCTION AT SOURCE

The requirement for efficient power conversion requires rapid switching of current and voltage by the main switching transistor. This gives rise to high $\dfrac{dI}{dt}$ and $\dfrac{dV}{dt}$ values which cause EMI. The major source of EMI in a switched mode power supply is around the main switching transistor. The abrupt transitions of switching current and voltage shock excite ringing or oscillations in the parasitic capacitances and inductance of transformers, chokes and wiring. In order to reduce this interference at source, a slight trade-off can be made with efficiency by deliberately reducing the speed of the current changes, $\dfrac{dI}{dt}$, and voltage changes, $\dfrac{dV}{dt}$, at the switching devices. This can be achieved by the addition of small series inductors to limit $\dfrac{dI}{dt}$ and small shunt capacitors to limit $\dfrac{dV}{dt}$ as shown below.

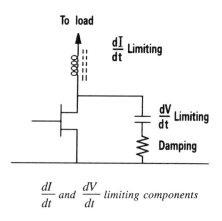

$\dfrac{dI}{dt}$ and $\dfrac{dV}{dt}$ limiting components

This solution will give a slight increase in the power dissipated in the switching device.

Another source of interference is the fast fall time current spikes which occur when diodes are reverse biased. Typically a fast recovery diode will "snap off" in around 10 ns. This can result in ringing and radiation in the very high frequency region.

The fast fall time current spikes arising from diode snap-off can be controlled by the use of "soft recovery" diodes or by the incorporation of a low value ceramic capacitor connected across the diode. A small trade-off can be made with efficiency by using small RF inductors in series with the main switching transistors to limit the switching edges, thereby giving a substantial reduction in EMI at source. Additionally the circuitry should be contained in a grounded metal screening enclosure in order to prevent radiated interference. Usually effective suppression is only achieved by using both filtering and screening together, as neither is fully effective on its own.

Wiring Layout

Another area that repays close attention for EMI reduction at the initial design stage is the physical layout of the supply. All leads carrying rapidly switched currents and voltages should be kept as short as possible to minimise radiation. Excessive wiring length increases inductance and capacitance, encouraging unwanted voltage drops. It is particularly important to pay special attention to the switching paths in the supply. Excessively long wiring can act as an aerial causing unwanted electric and magnetic fields, which easily couple into other wiring. Examples of good and bad wiring layouts are shown below. In addition, high current and high voltage wiring should be kept away from sensitive parts of the circuit such as the voltage reference and the low-level sense circuits in the regulator. In practice this will mean placing the main switching devices close to the transformer and choke, and also the provision of screening.

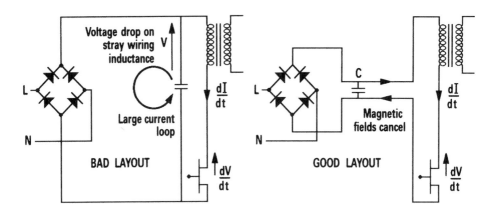

Examples of good and bad wiring layouts

Grounding

Grounding is an important aspect of interference control. There are two configurations commonly used for grounding within equipment. These are shown below.

BAD GROUNDING PRACTICE

Series connection - single point

GOOD GROUNDING PRACTICE

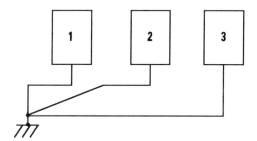

Parallel connection - single point

Two points should be borne in mind when discussing grounding:

• Separate ground points on a chassis are seldom at the same potential;

• All conductors and wires have a finite impedance.

Inclusion of these points gives a different appearance to the first of these two earth circuits.

BAD GROUNDING PRACTICE

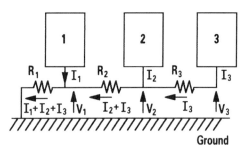

Ground

Series connection - single point, plus parasitic resistances

The first point dictates that a common single point earth is desirable; unfortunately in practice this tends to mean that a number of common impedance paths are introduced into the earth returns for units 1, 2 and 3. A situation where this occurs commonly is in printed circuit boards which pick up all their connections including ground through an edge connector. This can result in excessively noisy grounds especially if one of the stages is a high powered stage.

At low frequencies where wiring inductance is not a problem the solution is to use a single point grounding scheme, such as that shown overleaf.

GOOD GROUNDING PRACTICE

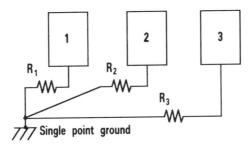

Parallel connection - single point, plus parasitic resistances

This scheme avoids the problems of common ground return paths and impedances; however, it is cumbersome and can necessitate long ground connections.

For frequencies above around 1MHz a low impedance ground plane is good practice especially on printed circuit boards. Here multi-point grounding is the preferred method, provided the ground connections are kept short.

The importance of choosing grounding points carefully can be illustrated in the following simple SMPS circuit. The gate of the transistor is driven from the secondary of a pulse transformer to provide mains isolation for the control electronics. In the first circuit the lower connection from the secondary of the gate drive transformer includes a length of common wire connection which carries the source current. At the switching frequencies in common use this wire could have a significant inductance causing a voltage drop to occur along its length. This voltage could easily be large enough to cause faulty turn-on cycles in the MOSFET. By making the transformer connection directly to the source of the transistor this problem can be avoided.

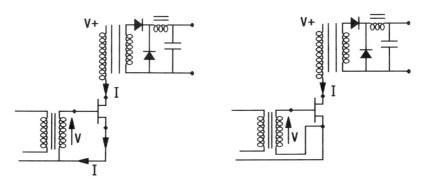

Method of avoiding faulty turn on cycles due to wiring inductance

In practice it is often impossible to stick rigorously to one grounding scheme or another and the following guidelines should prove useful.

- Never combine noisy, "high power" circuit grounds with ones for "low powered" signal circuitry.

- Group ground leads selectively, ie. place high power grounds together and keep low power grounds together.

- Provide separate grounds for chassis, case and a.c. power ground when required for safety.

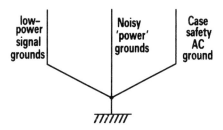

Grounding guidelines

A further point to watch out for in a design, when planning the earthing or grounding, is to avoid the unintentional inclusion of earth loops. These are closed circuit loops, often produced inadvertently, which act as pick-up loops for interfering signals induced by stray magnetic fields. If such loops are unavoidable their area should be minimised.

EMI FILTERS

Filtering of the input and output leads to contain EMI within an SMPS is an important aspect of interference control. The components available for this purpose are capacitors, inductors and feed-through capacitors and filters. These must be chosen with care and appropriate types picked for the frequencies at which the filtering is to be effective. The most important consideration for choosing a component for this application is the value of its "parasitics". Take, for example, the capacitor shown in the equivalent circuit below.

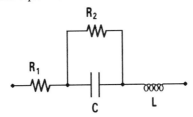

L&R are parasitic

Capacitor equivalent circuit

This has series inductance, L, arising from the leads and the physical construction of the component and resistances associated with its losses. It is important that the effects of the parasitics are small at the frequency of interest if the capacitor is to be effective at decoupling or bypassing.

Capacitors for EMI suppression and filtering should have good high frequency properties and a low series inductance. The most useful types are disc and plate ceramic components. These have excellent high frequency properties with good high voltage properties. It is important to consider the effect of lead length on the layout of the filter and the effectiveness of the bypass capacitors. The table below gives the self-resonant frequency for ceramic capacitors of different values with different lengths of leads.

Self-resonant frequency MHz		
capacitance pF	0.25in leads	0.5in leads
10,000	12	-
1,000	35	32
500	70	65
100	150	120
50	220	200
10	500	350

It is important to use capacitors whose self-resonant frequency is well above the frequencies at which bypassing or decoupling is to be effective.

At switching frequencies where a higher capacitance is required than is available from ceramic types, polycarbonate or polyproplyene types are the most effective.

Inductors

Inductors may be classified by their core type. The most general types are the air-cored varieties and the magnetic core types. Magnetic core types can be further subdivided depending on whether the core is open or closed. Designing EMI filters would not be a problem if the inductors were ideal. Unfortunately they have stray capacitance arising from adjacent turns and stray resistance from the wire of which it is wound.

In a low pass filter configuration the inductor will only be effective if it is operated well below its self-resonant frequency; thus, ideally, only the smallest inductance should be used that will give the necessary attenuation, at the lowest frequency at which the filter has to reject.

Another important consideration in the choice of an inductor is its stray magnetic field. Closed core types, eg. toroids, have a much reduced surrounding magnetic field as compared to air or the open cored types. Consequently they are much less prone to radiate or receive signals by means of stray magnetic coupling. If a toroidal inductor cannot be used it may be necessary to provide shielding by enclosing the winding in a grounded metal box.

The Bifilar Wound Choke

A useful component for filtering power supply lines between equipments is the bifilar wound choke. It is especially useful when common mode or ground noise is a problem and finds almost universal use in switched mode power supplies. It consists of bifilar winding placed on core to form a broadband transformer which allows equal and opposite currents to flow through its windings while suppressing unequal opposite currents such as ground noise. Because of the bifilar winding no net flux is set up in the core when the currents are balanced, therefore balanced currents do not encounter any inductance. When unbalanced currents flow there is a net flux giving an inductance and the choke opposes such currents. An additional advantage of the balanced configuration is that in normal operation the core is kept well away from saturation. The figure below illustrates the use of a bifilar choke between a supply and a load.

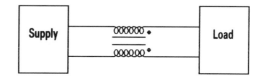

A bifilar choke used to block EMI between a supply and a load

Bifilar chokes are used to break ground loops as shown below. This figure shows an equivalent circuit of a bifilar choke including an interfering ground potential.

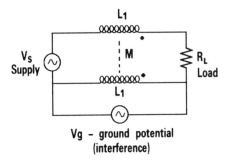

Equivalent circuit of a bifilar choke

For a balanced supply only (no ground noise), neglect V_g; the equivalent circuit becomes:

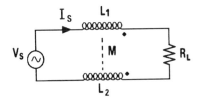

Equivalent circuit for a balanced supply only

$$V_s = j\omega\left(L_1 + L_2\right)I_s - 2j\omega MI_s + I_s R_L$$

where:

$$L_1 = L_2 = M$$

Therefore:

$$I_s = \frac{V_s}{R_L}$$

The balanced supply is unaffected by the choke.

For an unbalanced (ground) voltage the circuit becomes:

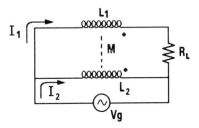

Equivalent circuit for an unbalanced supply only

For outer loop:

$$V_g = j\omega L_1 I_1 + j\omega M I_2 + I_1 R_L$$

For inner loop:

$$V_g = j\omega L_2 I_2 + j\omega M I_1$$

But:

$$L_1 = L_2 = M = L \quad (say)$$

Therefore:

$$I_1 R_L = 0$$

Therefore there is no current in R_L due to V_g.

Ferrite Beads

Ferrite beads are a simple and convenient way to increase the high frequency loss in a conductor, without affecting its low frequency and d.c. properties. They are small cylinders of a lossy ferrite material, with an axial hole through the centre. They can easily be slipped over the leads of components to provide attenuation of high frequency signals. They are most effective above 1MHz and when properly used can give effective high frequency decoupling and shielding. They are particularly useful in applications where a high current is flowing and it is not possible to insert a resistor in the circuit. Shown below is the equivalent circuit of a ferrite bead at high frequencies.

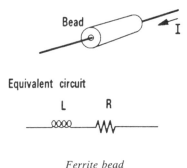

Ferrite bead

The resistive component arises from the loss component of the ferrite. Ferrite beads are most useful for damping out high frequency oscillations and "ringing" in switching circuits. They are also useful for blocking high frequency noise on power lines. The impedance of most beads is limited to around 100Ω, making them most effective in low impedance lines, such as power supplies and switching circuits. Typically, ferrite beads are used in conjunction with shunt capacitors to form low pass filters to block EMI on lines. Care should be taken in their use not to introduce any spurious resonances into a circuit which could exacerbate an interference problem. Another point that is often overlooked is that the ferrite must not be saturated.

Feed-Through Filters

Feed-through filters are useful components for EMI suppression. They consist of a central ferrite bead to supply series inductance between two coaxial ceramic capacitors formed from a high dielectric ceramic material. These are manufactured by several companies and are available in a variety of voltage and current ratings, covering a range of frequencies. They are available as both L and pi low-pass sections and are convenient to use. The figure below shows the construction of a typical feed-through filter.

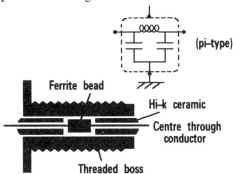

Feed-through filter construction and equivalent circuit

Input Filters

The interference problem is most severe at the input of a switched mode power supply. At the output there is generally a low pass filter section, to remove the switching frequency ripple, and this serves to attenuate high frequency noise. It is usually only necessary to add a relatively simple single section LC low pass RF filter on the output lines in order to meet most EMI output specifications.

The filtering at the input must be capable of blocking the switching frequency ripple and all of its harmonics well into the MHz region of the spectrum. This ensures that no interference is fed back into the prime power supply. Typically, to meet most input noise specifications a mains input filter such as that shown below is used.

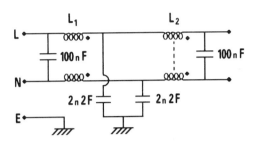

A typical mains input filter

In this input filter two bifilar, or common mode, chokes are used along with four bypass capacitors. The values of the chokes depend upon the switching frequency and the amount of attenuation required. The value of the ground bypass capacitors is limited by the amount of earth leakage current that is acceptable. Normally this is close to 2.2nF or 4.7nF.

At the output the filtering problem is normally easier as there is already a low-pass filter to smooth the output current. Typically, for a switching frequency of 40kHz, a single section low-pass filter comprising a 100µH inductor and a 47µF capacitor will be effective for blocking the switching frequency. At harmonics in the MHz range, where EMI is usually a problem, the stray inductance of the capacitor will make it ineffective as a bypassing component. The interturn capacitance of the choke will appear in parallel with its inductance and eventually may bypass it at high frequencies, providing very little attenuation of EMI. The effect of these parasitics is shown below.

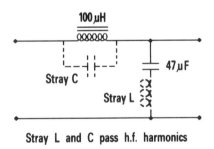

Stray L and C pass h.f. harmonics

A typical LC output filter showing strays

The solution to this problem is to use two filter sections, one to perform the low frequency filtering at the switching frequency and a second section to remove the high frequency components. The switching frequency filter is placed from the positive line to the negative line, while the EMI filter is placed from each line to ground as shown below. Typically, the EMI filter would use ceramic capacitors of values around 10-100nF. A small series inductor can be added if required. In this way the high frequency components are bypassed to ground, while the low frequencies are bypassed line to line.

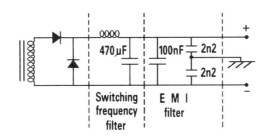

LC output filter with additional high frequency bypass components

Components for use in such filters should be chosen carefully. The inductors must be designed so that they are not magnetically saturated at the current levels drawn from the supply. The RF inductors are often wound on cores of a lossy material such as powdered iron to give them a dissipative component at high frequencies.

The capacitors bypass the unwanted RF noise to ground. These must be chosen so that they do not pass an excessive a.c. current from the supply as this constitutes an a.c. ground leakage path. Ceramic disc types are normally preferred in this application as they have a low series inductance and good RF properties.

EMI SCREENING

Another important consideration in the design of switched mode power supplies is the prevention of radiated interference from the supply entering into surrounding systems and wiring. This is achieved by paying careful attention to screening and grounding throughout a design. Switching regulators can generate noise up to frequencies in the VHF range. These frequency components are easily radiated from packages that are not RF tight, in other words the circuitry of the supply should be fully enclosed in a grounded, conducting case. In addition all leads, input, output and control functions, must be adequately filtered at the point of entry to the case. All lids and covers should be bonded onto the main chassis using RF gasket material. For optimum screening the enclosure should be thought of as being electrically water-tight.

Metal Screens

When planning screening for a supply it is necessary to know the type of field that is being shielded against. The characteristics of the field are determined by the nature of the source. A source with a high current and low voltage will have a near-field that is predominately magnetic while one that has a high voltage and low current has a predominately electric field. The electric and magnetic fields must be considered separately. Electric fields are relatively easy to shield against by using a metal sheet of good electrical conductivity, as the prime shielding mechanism is reflection.

Magnetic fields such as those around a transformer are more difficult to shield against. The prime mechanism for magnetic shielding is absorption loss, which is dependent on both the relative permeability and the conductivity of the shield. It would appear that a material with a high relative permeability will give the best results as a magnetic screen, but the permeability of most magnetic shielding materials falls off rapidly with increasing frequency. At the frequencies of interest in switched mode supplies it is probably just as effective to use sheet steel as a magnetic screen. Additionally, the magnetic properties of sheet steel do not degrade when it is worked in the way they do in high permeability materials such as mu-metal.

In practice it is difficult to fully enclose a unit such as a power supply in a metal box as some provision must be made for cooling either by convection or forced air cooling. This necessitates the provision of holes for the air to pass through.

The shielding effectiveness of a square grid of side l, of round holes of diameter d, and pitch c is

$$S = 20\log\frac{c^2 l}{d^3} + 32\frac{l}{d} + 3.8 \ dB$$

where S, the shielding effectiveness, is the improvement in shielding over a square hole of side l. This assumes that the holes are acting as short waveguides above cut-off; ie. $d < \frac{\lambda}{2\pi}$, where λ is the wavelength at the frequency of interest.

Other areas to pay particular attention to in the design of screens are the joints and seams in cases and also the seating and electrical bonding of lids and covers. If a lid or joint is not tight fitting it is relatively easy for it to act as a slot antenna, making it an excellent radiator and receiver of interference. To avoid this, the use of conducting gaskets is recommended in all joints and covers.

Electrostatic Screens in Transformers

One point that is often overlooked is the capacitive coupling of noise across the transformer from the primary to the secondary or vice-versa. The effect of this coupling across the transformer is to produce the ground currents shown. The return path for these is through mains supply and the 4.7nf capacitors. Ideally the current in the mains-to-ground impedance should be kept to a minimum. As the value of the filter capacitors is limited by safety considerations the only remaining option is to reduce the ground current itself.

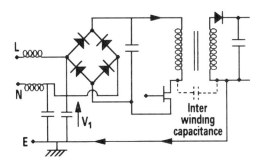

Ground currents caused by transformer inter-winding capacitance

This can be achieved by incorporating a Faraday screen into the transformer and connecting it back to the primary circuit. In this way the ground noise circuit path is broken and the noise current is returned to the primary as shown below.

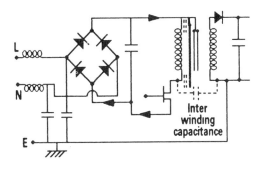

Transformer screen used to break ground current path

The screen is implemented by winding a layer of copper tape between the primary and secondary in such a way that it does not provide a "shorted turn". The copper tape has no effect on the magnetic coupling between the primary and the secondary.

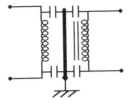

Use of screen between primary and secondary of transformer

Another source of unwanted noise can arise from currents coupled to ground through the transformer core if it is clamped to the chassis ground. The solution is to electrically connect the core to the primary side of the transformer.

Switching Transistor Mounting and Heatsinking

The primary noise sources in a power supply are the power switching transistors. Unfortunately they must be mounted in such a way as to dissipate their heat into the chassis, which usually sets up ground currents. A way must be employed to mount them so that they are free to dissipate heat and not induce ground currents. The capacitance of a TO-3 transistor package mounted on a heatsink with a mica washer is typically of the order of 100pF. Given a 200 volt input with almost perfect switching at 20kHz, over 1mA at 1MHz could flow in the ground. The ground current path is shown below.

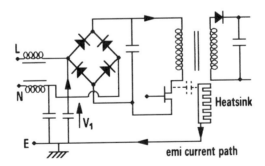

Ground currents arising from heatsink capacitance

The solution is to mount an electrical screen, between the transistor and the heatsink, which is electrically insulated from both, but allows the passage of heat. This screen is connected back to the prime power supply line, rather than ground.

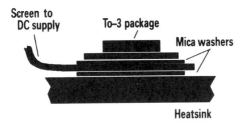

Transistor heatsink screen

This provides a return circuit for the noise other than through the chassis. The ground current path is shown below, along with a typical mounting arrangement.

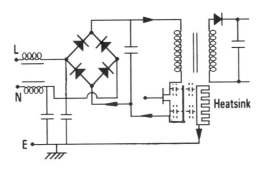

The use of transistor heatsink screen to break ground current path

Screening Guidelines

The following points should be borne in mind when planning enclosures and shields.

- Magnetic fields are harder to shield against than electric fields.

- Use a good conductor against electric fields, plane waves and high-frequency magnetic fields.

- Actual screening effectiveness obtained in practice is usually determined by the leakage at the seams and joints, not by the material itself.

- The maximum dimension (not area) of a hole or gap determines the amount of leakage, eg. slot radiator.

- A large number of small holes results in less leakage than a larger hole of the same total area.

EMI MEASUREMENT AND SPECIFICATIONS

There are two main EMI specifications that must be checked out, that for conducted noise and that for radiated noise. Usually if a design includes adequate screening and the conducted noise is kept under control, by means of appropriate filters, radiated noise is not too much of a problem.

Conducted Noise

The set-up used for measuring conducted noise is shown below. The SMPS is powered from the mains via a Line Impedance Stabilisation Network (LISN). The function of this circuit is to connect a specified rf terminating resistance of 50Ω across each mains input terminal of the SMPS to ground, over which the rf voltage from the power supply can be measured. This rf voltage is measured with a calibrated receiver or a spectrum analyser with a 50Ω input impedance and a closely specified bandwidth. The LISN can be switched to measure the rf voltage on the line or the neutral connection.

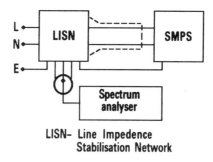

LISN- Line Impedence
Stabilisation Network

A test setup for measuring conducted noise

The figure below shows a simplified schematic circuit for a LISN.

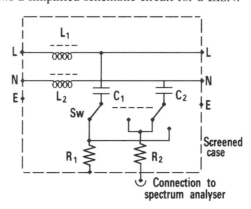

Simple schematic for a LISN

There are different national and international organisations that specify acceptable interference limits from electronic equipment including switched mode power supplies. Generally these define limits for both conducted and radiated interference. In the UK BS800 applies, in the US the FCC (Federal Communications Commission) rules apply, in Germany VDE0871 and internationally CISPR (Comité International Special des Perturbations Radioélectrique) apply. The limits for conducted interference on the input and output leads of a power supply are shown below. Broadly the different national limits are very similar.

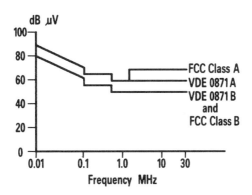

The different limits for conducted EMI

The vertical scale is in dBμV, that is decibels above 1μV, so 60dBμV is 1mV. As mentioned previously there are some minor variations in the different national specifications but a power supply is acceptable in all countries if the rf voltage at its input terminals at frequencies above 150kHz is less than +54dBμV or 500μV.

Radiated Interference

Radiated interference is normally specified in terms of the acceptable maximum field strength at a given distance from the equipment. Measurements are normally carried out using a testing range known to be free from reflections and extraneous signals. The power supply is set up in normal operation and a spectrum analyser or tunable receiver is used with a calibrated aerial to measure the electromagnetic field strength at different frequencies. Shown overleaf is a typical radiated noise measurement set-up.

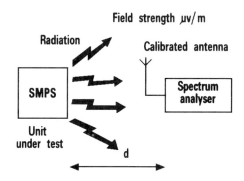

The setup for measuring radiated EMI

The radiated noise limits for VDE0871 are given in the following table.

RADIATED-VDE CLASS B LIMITS

Frequency Range	Recommended Bandwidth	Level	Recommended Sensor/Antenna
10kHz to 150kHz	200 Hz	50µV/m at 30m	Magnetic loop 60cm per side maximum
150kHz to 30 MHz	9 kHz	50µV/m at 30m	Magnetic loop as above or 1m vertical rod
30 MHz to 470 MHz	120 kHz	50µV/m at 10m	Balanced dipole tuned to 80MHz or higher,1-4m height above ref. plane
470 MHz to 1 GHz	120 kHz	200µ V/m at 10m	Balanced dipole-plane polarized Directional-plane polarized antennas may be used

EMI field strength limits specified in VDE0871 class B

SUMMARY

The following are some points to consider at the design stages of a switch mode power supply to help with the reduction of EMI.

- Consider a slight trade-off of switching speed and hence efficiency for an easier EMI problem!

- Plan on a fully encased, grounded enclosure to screen and house the supply.

- Keep high $\dfrac{dI}{dt}$ and $\dfrac{dV}{dt}$ lines as short as possible to reduce the noise radiation.

- Keep the input and output leads as far as possible from the electric and magnetic noise generators.

- Keep switching paths simple, to prevent the creation of excessive ground paths.

- Be careful with grounding.

- Provide shielding between the noise sources and sensitive input and output circuits, including sensitive voltage references.

- Minimise capacitive coupling into the chassis.

- Provide effective rf filtering on all input and output connections from the screened enclosure.

- Consider a Faraday screen in the main transformer.

4.2   PROTECTION OF SMPS

In the design of a power supply it is prudent to provide protection circuitry to protect against extreme and abnormal operating conditions that will inevitably occur when the supply is in use. These can occur in the form of output short circuits and excessive loads or high voltage transients on the input supply line. Many of the components in a power supply are handling powers greatly in excess of their dissipation capability. Under fault conditions it is quite possible that they may start to dissipate this power, leading to their rapid failure. The power supply designer has no control over these faults and therefore must incorporate circuitry to accommodate them safely. This falls into four broad categories:

- overcurrent protection

- overvoltage protection

- inrush protection

- device protection and snubber circuits

OVERCURRENT PROTECTION

In order to provide current-limiting, some means of sensing overcurrent conditions must be provided.

The normal method of providing current-limiting in an inverter or power supply regulator is to provide an output current sensor driving a feedback path that is used to control the output current. Below a critical threshold level the feedback does not affect the output current but above the threshold a high resistance path is introduced into the circuit to control the current. In a switched mode power supply the commonest method of achieving current limiting is to control the mark-space ratio of the switching transistors. Under fault conditions the transistors can be switched off.

The purpose of current-limiting is two-fold: firstly to limit the dissipation in the regulator components to safe values and thereby prevent damage to them, and secondly to provide some protection to circuits and systems being powered by the supply.

Three types of current-limiting are normally used. Firstly, the straightforward current limiter merely prevents the current rising above a critical value by dropping the output voltage to zero. This characteristic is shown below.

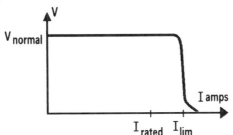

A typical current limiting V-I characteristic

This is relatively simple to implement but can have the disadvantage that under overcurrent conditions it can still over dissipate the regulator device. A second, more effective, solution is to use a foldback current limiter with the current voltage characteristic below.

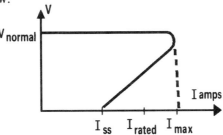

A typical foldback current limiting V-I characteristic

Thirdly, cycle-by-cycle protection is a useful method of output current limiting in an SMPS. It is usually implemented by means of a current transformer placed in the primary circuit of the main switching transformer. In this way a signal is fed back to the controller circuit which is used to reduce the pulse width under overcurrent conditions. Under short circuit conditions the output can be switched off completely. It has the advantage of fast operation on a cycle-by-cycle basis and can be implemented with few extra components. Most SMPS controller ICs incorporate it as a standard facility.

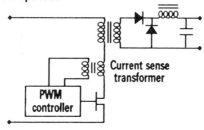

Cycle-by-cycle current limiting

OVERVOLTAGE PROTECTION

Overvoltage protection must deal with three possible situations:

- Reverse voltage on output

- External overvoltage on output

- Internally generated overvoltage

The first two situations are reasonably easily dealt with by placing "catcher" diodes on the output. For reverse polarity protection, a normally reverse biased diode can be placed on the output. Normal overvoltage protection can be provided by a zener or avalanche diode whose voltage is in excess of the normal operating voltage of the power supply. Diodes for both types of protection must be amply rated to cope with the anticipated fault conditions.

It is important that a power supply does not give out an abnormally high voltage under fault conditions. If it did so it could easily damage the circuitry which it is powering. There are two widely used schemes for overvoltage protection in SMPS, namely inverter-inhibiting protection and crowbar protection.

Turning first of all to inverter-inhibiting protection, a block diagram of this scheme is shown below.

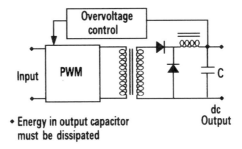

Inverter inhibiting over-voltage protection

The d.c. output voltage is monitored by the overvoltage control circuit. This is usually some form of comparator which is set to trigger under overvoltage conditions. Its output is used to switch off the pulse width modulator (PWM), thereby shutting down the power supply, in much the same way as was done for cycle-by-cycle current limiting. One problem with this system of overvoltage protection is that the energy stored in the filter capacitor must be dumped into the load, possibly causing damage.

To get round this problem we can use the "crowbar" circuit shown below.

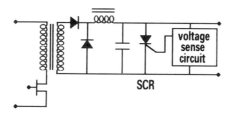

Crowbar overvoltage protection

This can be made to shut down the power supply very quickly, within a few microseconds. It operates by sensing the output voltage and when an overvoltage occurs, the silicon control rectifier (SCR) is fired, shorting the output. This scheme depends upon the power supply having adequate current limiting and the SCR being rated to withstand the short-circuit current along with the discharge surge from the output filter capacitor.

INRUSH PROTECTION

The input filter smoothing capacitor used in many switched mode power supplies causes a potential problem. Shown below is a typical input circuit of an SMPS. Included in the circuit are the parasitics associated with the capacitor and the wiring.

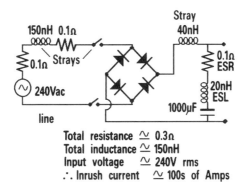

Typical input circuit showing parasitic inductances and resistances

The filter capacitor is deliberately chosen to have a high capacity and a low equivalent series resistance (ESR). On application of power it appears very close to a perfect short circuit, causing a very high current surge until it is charged. These current surges can exceed the ratings of the diodes in the input rectifier by many times, destroying them. This inrush current surge can be of very short duration and hence may not contain enough energy to open circuit breakers or blow fuses. The exact amplitude of the surge will vary depending on when in the input voltage cycle the input switch is closed; however, it is necessary to design for the worst case when the switch closes at the maximum of the voltage waveform. The magnitude of an inrush surge is only limited by the resistance and inductance of the wiring associated with the input circuit and the ESR of the capacitor. Typically, inrush currents can reach many hundreds of amperes.

Two methods of inrush protection are shown below. The first circuit uses a resistor to limit the peak surge current. It has the advantage of simplicity but suffers from the drawback that it reduces efficiency due to losses in the resistor.

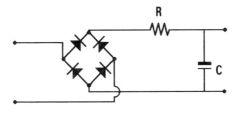

A simple resistive surge limiter

An alternative is to use a thermistor. This component has a high resistance when cold and a low resistance when hot.

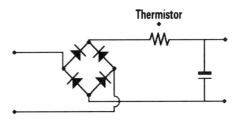

A thermistor surge limiter

On switch-on the thermistor is cold and in its high resistance state, thereby limiting the initial current. The current causes the thermistor to self-heat, reducing its resistance once the inrush period is over. This overcomes most of the efficiency problems associated with the resistor method, but gives rise to new problems when "dropouts" occur on the prime power supply, caused by the thermal time constant of the thermistor. During a dropout the thermistor may not get sufficient time to cool to its high resistance state and will not afford protection to the circuit when the supply comes on again.

A better solution is to use a resistor that is switched out of circuit as the voltage across the capacitor builds up. The classic method of doing this is with a relay whose coil senses the capacitor voltage. A faster acting solution is to use a silicon controlled rectifier (SCR) or a TRIAC to switch out the resistor. The circuit below shows such a scheme.

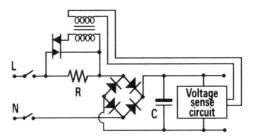

Triac switched inrush protection

The circuit chosen will depend on the type of prime power supply in use, the degree of protection required for the input circuitry and the overhead of including the additional circuitry. It is worth remembering that the inclusion of effective protection will contribute to the reliable operation of the unit.

SNUBBERS AND DEVICE PROTECTION

Turn-Off Snubber Circuits

Snubber circuits are used to reduce the losses in switching transistors by altering the waveforms around them. To understand the need for a snubber consider the simple circuit shown below. When the transistor is turned off the collector voltage will rise instantaneously, but the collector current will flow until the energy stored in the leakage inductance of the transformer has been discharged. This means that power is dissipated in the transistor at switch-off.

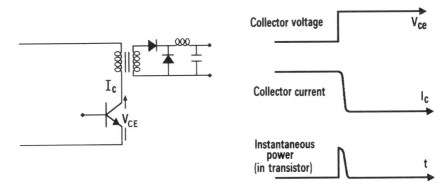

Transistor switching circuit without turn-off snubber

If a snubber circuit is added the transistor, to slow the rise in collector voltage, the power dissipated by the transistor is reduced.

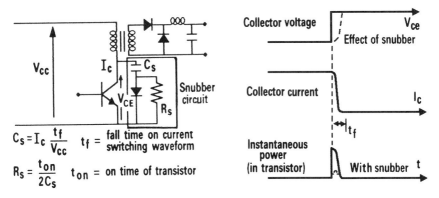

$$C_s = I_c \frac{t_f}{V_{cc}} \qquad t_f = \begin{array}{l}\text{fall time on current}\\ \text{switching waveform}\end{array}$$

$$R_s = \frac{t_{on}}{2C_s} \qquad t_{on} = \text{on time of transistor}$$

Transistor switching circuit with turn-off snubber

The operation of the snubber is as follows. Capacitor C_s holds V_{ce} low during current turn-off, as the diode is forward biased. Resistor R_s is chosen to discharge C_s in less than the minimum on-time, t_{on}. Hence the time constant $R_s C_s$ is chosen to be a half of t_{on}.

$$C_s = i_c \frac{t_f}{V_{ce}}$$

$$R_s = \frac{t_{on}}{2C_s}$$

where t_f is the device switching time.

Turn-On Snubber Circuits

Again, the turn-on snubber is used to limit the power dissipation during switching. It is mainly useful when the transformer leakage inductance does not provide sufficient delay between the fall in V_{ce} and I_c at turn-on.

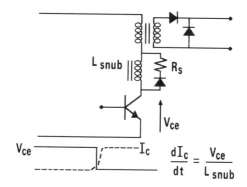

Turn-on snubber circuit

The circuit shows a typical turn-on snubber circuit. During turn-on the series inductor serves to isolate the transistor from the load, permitting the transistor to turn on before it passes significant current. The rate of rise of the collector current $\frac{di_c}{dt} = \frac{E_{in}}{L_{snubber}}$. The resistor diode network around the snubber inductance provides a current discharge path to dissipate the energy stored in the inductor during the off period.

It is worth noting that snubber circuits do not reduce the overall dissipation in a power supply, but do reduce the dissipation of the switching device. They work by dissipating the switching energy in the snubber resistor rather than the device.

Device Protection

While dealing with protection it is probably worth mentioning voltage clamping. Here an avalanche diode is placed across the collector and the emitter of the transistor. The breakdown voltage of the diode, V_z, is chosen so that it is below the collector-emitter breakdown voltage and greater than the supply, V_{supply}. In normal operation the diode will not conduct, but it will serve to clip any high voltage transients which might be harmful to the transistor.

Voltage clamping

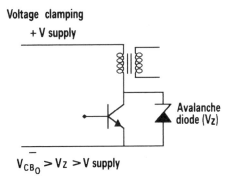

$\overline{V_{CB_0}} > V_Z > V$ supply

Switching device protection

The protection measures outlined in this section will considerably increase the complexity of a SMPS but they are well worth considering as they will greatly enhance its reliability.

4.3 RELIABILITY AND COOLING

For a power supply to be useful it must be reliable. The overall reliability of a system is strongly dependent on the reliability of its power supply. Usually a system can incorporate some degree of redundancy to enhance reliability but it is more difficult to incorporate redundant power supplies.

Reliability is a measure of the ability of a component or piece of equipment to perform its function, in an adequate manner, for a stated period of time. It is a necessary attribute for a product to be fit for the purpose for which it is designed and every bit as important as the actual functionality of the product.

The reliability of a component or system is only definitely known once it has failed. For a knowledge of reliability to be useful it must be possible to make predictions of the expected life in order to minimise the incidence of failures in service. Reliability, in an engineering sense, becomes the prediction of life of a component or system based on knowledge of the previous failure history. Reliability is a "state of knowledge" rather than a "state of things".

In order to be able to predict the reliability of a component it is necessary to know its statistical failure rate. This knowledge is based on data gathered from either past operational experience of the component or from batch testing. For most most components the failure rate varies with time according to the well known "bath-tub" curve shown below.

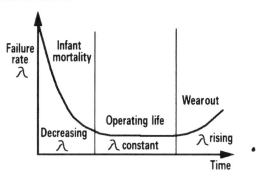

The bath-tub curve

During the early stages of life (region 1) the failure rate is high and decreasing. This is often referred to as the infant mortality period where components with inherent manufacturing weaknesses will fail prematurely. During the middle life period (region 2) the failure rate is constant and caused by random failures. This period is the useful life region. Finally the wearout region is reached where failures occur because of aging effects and the failure rate starts to slowly increase. In normal operation semiconductor components do not generally reach this phase of operation. In practice manufacturers try to make the infant mortality period as short as possible by careful quality control and process control during production.

Additionally parts may be subjected to a "burn-in" period before dispatch to weed out potential failures and they are only released when they are in their useful life period.

Quantifying Reliability

Various measures are used to quantify reliability, the commonest being failure rate,λ, and the mean time to failure, MTTF. These are related thus:

$$MTTF = \frac{1}{\lambda}$$

Typical values of failure rate for a high reliability part are around 10^{-8} failures per hour. Another measure often used is the mean time between failures, MTBF. Strictly this is the mean time between failures as the device goes through successive cycles of failure and repair and is related to the MTTF thus:

$$MTBF = MTTF + repair\ time$$

For semiconductor parts repair is not normally possible and the MTTF and MTBF are often used interchangeably.

The reliability, R(t), of a component is related to its failure rate and time, t, thus:

$$R(t) = e^{\left(-\lambda t\right)}$$

This assumes that the failure rate is constant and random. This assumption can be justified because of the many diverse forces leading to failure for complex parts. These conspire to produce, in effect, random failures.

To calculate the reliability of a system comprising many components the individual failure rates are summed and used to calculate the overall reliability, thus:

$$\lambda_{total} = \lambda_1 + \lambda_2 + \ldots\ldots + \lambda_n$$

$$R_{total}\left(t\right) = e^{-\left(\lambda_{total}t\right)}$$

The degradation of semiconductors usually results from chemical reactions and physical processes which change the structure of devices on an atomic scale. The rate at which such reactions occur, and hence the degradation of the reliability with temperature, is well described by the Arrhenius equation:

$$R = R_o e^{\left(\frac{-\Delta E}{KT}\right)}$$

where

R = Reliability as a function of temperature

R_o = constant

ΔE = Activation energy

K = Boltzmann's constant

T = Temperature in degrees Kelvin

Typical values of activation energy, ΔE, are around 0.75ev; thus a 10 degree temperature rise will halve reliability.

Use can be made of this to overcome the infant mortality period (in the bath-tub curve) by raising the operating temperature to accelerate the failure rate. This process is often referred to as "burn-in". In this way manufacturing faults leading to infant mortality can be overcome quickly and economically, before products are sold.

RELIABILITY PREDICTION

Methods of reliability calculation are outlined in the US Department of Defense MIL-HDBK-217D. This contains failure rate data on all types of components that have been cleared for US military use. These data have been built up from experience of component failures and also from the results of accelerated testing at elevated stress.

In the early stages of a project, before detailed knowledge of the details of a circuit design is available reliability is estimated by means of the Parts Count Method.

Parts Count Reliability Estimation

This method is useful before detailed information regarding the operating stresses of individual components is available. The information required for this method is as follows:

- Generic part types

- Numbers of each part type

- Part quality level

- Equipment environment

The overall failure rate for the equipment is calculated using the relation:

$$\lambda_{EQUIPMENT} = \sum_{i=1}^{n} N_i \left(\lambda_G \pi_Q \right)_i$$

where:

$\lambda_{EQUIPMENT}$ is the total equipment failure rate

λ_G is the generic failure rate for the ith generic part

π_Q is the quality level of the generic part

N_i is the number of the ith part type

n is the number of part types

These values are tabulated in MIL-HDBK-217D. This method requires only a minimum of information, as no details of component operating conditions are needed. It is intended to give a reasonable estimate of expected reliability at the planning stages of a project.

Part Stress Reliability Estimation

As a design progresses more detailed design information is available and it is possible to estimate overall reliability by calculating the individual component failure rates. This method is known as Part Stress Analysis as it is based on the operating stress of each component.

A variety of different models are used depending on the component type; take, for example, discrete transistors and diodes:

$$\lambda_P = \lambda_b \left(\pi_E \times \pi_A \times \pi_Q \times \pi_R \times \pi_{S2} \times \pi_C \right)$$

where:

λ_P is the part failure rate

λ_b is the base failure rate

π_E is the environmental factor

π_A is the application factor (linear or switched)

π_Q is the quality factor (eg. JAN, JANTX etc.)

π_R is the power rating

π_{S2} is the voltage stress factor

π_C is the complexity factor (eg. dual, multiple emitter etc.)

A Cautionary Note

The use of the standard reliability prediction methods such as MIL-HDBK-217D can predict a higher MTTF than can actually be attained. MIL-HDBK-217D considers only static conditions and does not allow for transient conditions such as surges and spikes. Conversely, if a designer adds surge limiters, snubbers, soft-start circuitry, transient limiters and other protection devices, in order to obtain high actual reliability, MIL-HDBK-217D calculations will yield a lower MTTF, because of the extra complexity, than is attained in practice.

Be cautious of simple and inexpensive switch mode power supply designs with little protection and a high predicted MTTF!

Component Approval Systems

In order to build systems which have high reliability, which can be predicted with some confidence, it is necessary to use components of "approved quality". Several approval systems are in use and the choice will depend on particular customer requirements. One of the most widely used systems in the UK is the BS9000 system. It was established in the late 1960s for the independent inspection, approval and surveillance of manufacturers, distributors and test laboratories in the electronic component industry. For products destined for the Western European market components approved to CECC quality levels should be used.

The BS9000 and CECC specification system consists of generic specifications which relate to families of components; for example, fixed resistors or semiconductor devices. They contain the terms, definitions, standard values for characteristics, test methods and information relating to quality assessment procedures.

COOLING CONSIDERATIONS

It was stated in the previous section that the temperature of a semiconductor device strongly influences its reliability. In particular the temperature of the actual semiconductor chip, rather than the package is the controlling factor. To maximise reliability it is important to keep the device junction temperature as low as is reasonably possible.

The junction temperature, T_J, depends on the ambient temperature, T_A, the average power dissipation, P, and the junction to ambient thermal resistance, $R_{\Theta JA}$ (C/W), in the following manner:

$$T_J = T_A + PR_{\Theta JA}$$

This simple "d.c." thermal model is found to work well for switching power supplies since the waveforms are periodic and have a period much less than the thermal time constants involved.

The power dissipation is composed of three components in a switching regulator:

$$P = P_{blocking} + P_{conducting} + P_{switching}$$

Usually the power dissipated in the blocking state is negligible.

The junction to ambient thermal resistance can be considered to consist of three separate components shown below.

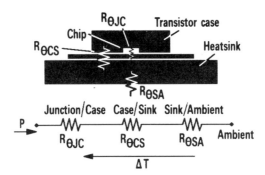

The junction to ambient thermal resistance of a transistor on a heatsink

where $R_{\Theta JC}$ is the junction to case thermal resistance, $R_{\Theta CS}$ is the case to heatsink thermal resistance and $R_{\Theta SA}$ is the heatsink to ambient thermal resistance.

$$R_{\Theta JA} = R_{\Theta JC} + R_{\Theta CS} + R_{\Theta SA}$$

The value of $R_{\Theta JC}$ is dependent on the internal structure of the transistor and its package and is determined by the semiconductor manufacturer. $R_{\Theta CS}$ and $R_{\Theta SA}$ are both determined by the mounting of the transistor. $R_{\Theta CS}$ is kept small by mounting the device using a heat conducting paste. $R_{\theta SA}$ is determined by the size of the heatsink in use and depends on the surface area to volume ratio. For most materials used as heatsinks, such as aluminium extrusion the thermal resistance per unit surface area is around 800°C/watt/cm². Hence the thermal resistance is given by:

$$R_{\theta SA} = \frac{800}{surface\ area} \quad °C/watt$$

The thermal conductivity of a heatsink can be improved by forced air cooling, and the graph below shows the thermal resistance versus volume for different air flow rates. The surface to volume ratio for extruded heatsinks is relatively constant.

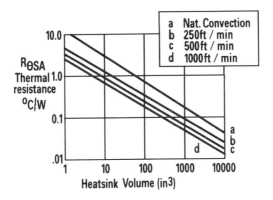

The thermal resistance of a heatsink as a function of volume and airflow

When specifying power devices caution should be exercised over the selection of one with a suitable dissipation rating. Most manufacturers specify their devices at 25°C case temperature. In most situations where a device is dissipating a significant power it is not possible to keep the case at 25°C and devices should be derated according to the thermal resistance of the heatsink and mounting arrangements. Allowance must also be made for the internal ambient temperature of the equipment.

4.4 CONCLUSIONS

A switched mode power supply designer requires a wide range of electronic skills, which most design engineers, individually, do not possess. These range from a knowledge of the different SMPS topologies, through wound component design, control theory, to RF and electromagnetic interference suppression.

Switched mode power supply design is a fast moving field, with an ever increasing proportion of electronic equipment using switching power supplies. One of the main driving forces in their development is the reduction of size and weight. To this end there are advantages to be gained from going to higher switching frequencies, in the high hundreds of kHz region, to reduce the wound component sizes. In addition circuit topologies using resonant power conversion can get round the problem of switching losses. Further reduction in volume still can be obtained by using novel wound component construction techniques and by the use of surface mount component technology.

The object of this book has been to introduce the problems that will be encountered when a switched mode power supply design is tackled. It is hoped that it will serve as a starting point for designers setting out to do a design and allow them to tackle it with some confidence.

INDEX

1-28-94

 ENGINEERING